POWER ELECTRONIC SYSTEMS
WALSH ANALYSIS WITH MATLAB®

POWER ELECTRONIC SYSTEMS
WALSH ANALYSIS WITH MATLAB®

Anish Deb
Suchismita Ghosh

CRC Press
Taylor & Francis Group
Boca Raton   London   New York

CRC Press is an imprint of the
Taylor & Francis Group, an **informa** business

MATLAB® is a trademark of The MathWorks, Inc. and is used with permission. The MathWorks does not warrant the accuracy of the text or exercises in this book. This book's use or discussion of MAT-LAB® software or related products does not constitute endorsement or sponsorship by The MathWorks of a particular pedagogical approach or particular use of the MATLAB® software.

CRC Press
Taylor & Francis Group
6000 Broken Sound Parkway NW, Suite 300
Boca Raton, FL 33487-2742

First issued in paperback 2017

© 2014 by Taylor & Francis Group, LLC
CRC Press is an imprint of Taylor & Francis Group, an Informa business

No claim to original U.S. Government works

Version Date: 20140311

ISBN 13: 978-1-4822-1596-0 (hbk)
ISBN 13: 978-1-138-07502-3 (pbk)

Visit the Taylor & Francis Web site at
http://www.taylorandfrancis.com

and the CRC Press Web site at
http://www.crcpress.com

To our families

For their patience and understanding during preparation of this book

Contents

List of Principal Symbols

ψ_n	$(n + 1)$th block pulse function
$\Psi_{(m)}$	block pulse vector of order m having m component block pulse functions
ϕ_n	$(n + 1)$th Walsh function in dyadic order
$\Phi_{(m)}$	Walsh vector of order m having m component Walsh functions
α	ratio of time period T_R and time constant T_L, triggering angle
β	extinction angle
γ	conduction angle
δ	logic operator
λ	scaling constant
τ_{av}	average developed torque
τ	instantaneous developed torque
τ_L	load torque
φ	phase angle
χ	Haar vector
ω	angular frequency, angular speed
A	pulse amplitude
B	coefficient of viscous friction
c_n	coefficient of the nth Walsh function
$c_{n\lambda}$	coefficient of the nth Walsh function in the scaled domain
c_n'	coefficient of the nth block pulse function
$\mathbf{C}\Phi_{(m)}(t)$	system output in Walsh domain
$\mathbf{C}'\Psi_{(m)}(t)$	system output in block pulse domain
$\mathbf{D}_{(m)}$	Walsh operational matrix of order $(m \times m)$ for integer differentiation
$\mathbf{D}_{\lambda(m)}$	Walsh operational matrix of order $(m \times m)$ for integer differentiation in the scaled domain
E	applied voltage
E_a	average applied voltage
f	frequency
$f(t)$	function of time
$\hat{f}(t)$	function represented via direct expansion in Walsh or block pulse function domain
$\bar{f}(t)$	function represented using operational matrix in Walsh or block pulse function domain
$\mathbf{G}_{(m)}$	Walsh operational matrix of order $(m \times m)$ for integer integration
$\mathbf{G}_{\lambda(m)}$	Walsh operational matrix of order $(m \times m)$ for integer integration in the scaled domain
$G(s)$	transfer function in Laplace domain
i_a	instantaneous motor armature current
I_a	average motor armature current

I_{ar}	rms motor armature current
I_F	normalized freewheeling current
I_L	normalized load current
I_{s1}	normalized current in the main thyristor
J	moment of inertia
K_a	DC machine constant related to armature
K_{af}	DC machine constant, product of K_a and K_f
K_D	Duty cycle of chopper
K_f	DC machine constant related to field
K_F	freewheeling cycle of chopper
K_{res}	DC machine constant related to residual magnetism
L_a	motor armature lumped inductance
L_L	load inductance
n	instantaneous motor speed
N	average motor speed
PCM	phase control matrix
PWMM	pulse-width modulation matrix
$\mathbf{R}\,\Phi_{(m)}(t)$	system input in Walsh domain
$\mathbf{R}'\,\Psi_{(m)}(t)$	system input in block pulse domain
R_a	motor armature lumped resistance
R_L	load resistance
SPMM	single-pulse modulation matrix
T	time period under consideration
T_L	load time constant
T_m	mechanical time constant
T_R	repetition rate of input waveform
THD	total harmonic distortion
v_o	load voltage
$\mathbf{W}_{(m)}$	Walsh matrix of order m
WOTF	Walsh operational transfer function
x	fraction lying between 0 and 1
z	sequency
Z_L	load impedance

Preface

About 80 years ago, Walsh functions were presented in the literature by Professor J. L. Walsh. However, he was not the first to propose a complete orthonormal function set that was very much unlike the well-known sine–cosine functions. Precisely, the Walsh functions were piecewise constant in nature as well as bivalued—and hence digital application friendly. Although Walsh functions are so different from the sine–cosine set, they still have some basic similarities with the good old sine–cosine functions. These fundamental properties made Walsh functions to become a strong candidate to application engineers from the mid-1960s.

The proposition of piecewise constant orthonormal function sets was pioneered by Alfred Haar in 1910. Haar function set is now known to be the earliest wavelet function having both scaling and shifting properties. Although each component function in the Haar set is bivalued, its amplitude is different from most of the other members. But in the Walsh set, all the component functions are piecewise constant and switch between two fixed values +1 and −1. This property seemed to be an added advantage for engineering application though the Walsh function set, Haar function set, and block pulse function set are related to one another by similarity transformation.

From the early 1970s, a horde of researchers started working with Walsh functions. They set the ball rolling with the solution of differential as well as integral equations, thus leading to many applications in the area of control theory, including system analysis and identification. Concurrently, the theoretical basis of Walsh analysis has also been strengthened mainly by control engineers and mathematicians.

Walsh functions first attracted Anish Deb (the first author) way back in 1982, when he noted the striking similarity between the shapes of different Walsh functions and different power electronic waveforms, i.e. the output waveforms of chopper, converters, inverters, etc. He initiated the idea of mingling power electronics with Walsh functions and making good use of the advantages offered by this "alternative" piecewise constant function set.

In this book, which is essentially a sort of "marriage" between power electronics and Walsh functions, we have explored many advantages offered by Walsh domain analysis of power electronic systems and have proposed a strong case in its favor to establish its right as an interesting as well as a powerful analysis tool for the study of power electronic systems.

The book starts with the background and evolution of power electronics, then proceeds gradually with a discussion of Walsh and related orthogonal basis functions and develops the mathematical foundation of Walsh analysis and first- and second-order system analyses by Walsh technique.

After presenting the underlying principles of Walsh analysis, the book deals with pulse-width modulated chopper, phase-controlled rectifiers, and inverter systems with many illustrative examples. The two appendices at the end of the book include a basic introduction to linear algebra and a few MATLAB® programs for some important numerical experiments treated in the book.

The book is targeted at postgraduate students, researchers, and academicians in the area of power electronics as well as systems and control. It may prove to be a source of new knowledge, and Walsh analysis hopefully deserves to be a new potential tool for further application and exploration.

Anish Deb gratefully acknowledges the support of the University of Calcutta in all phases of preparation of the book, while Suchismita Ghosh remains grateful to MCKV Institute of Engineering for providing her the opportunity for research and study to offer strong regular support to Deb.

Finally, the authors are indebted to CRC Press for accepting and publishing their work in such an attractive form.

MATLAB® is a registered trademark of the MathWorks, Inc. For product information, please contact:

The MathWorks, Inc.
3 Apple Hill Drive
Natick, MA 01760-2098 USA
Tel: +1 508 647 7000
Fax: +1 508-647-7001
E-mail: info@mathworks.com
Web: www.mathworks.com

Authors

Anish Deb, born in 1951, obtained his BTech in 1974 (recipient of Calcutta University silver medal), MTech in 1976 (recipient of two gold medals—Calcutta University gold medal and P. N. Ghosh memorial gold medal), and PhD (Tech) in 1990 from the Department of Applied Physics, University of Calcutta, India. In 1978, he started his career as a design engineer in a consultancy firm dealing with thermal power plant control and instrumentation and joined the Department of Applied Physics, University of Calcutta, as Lecturer in 1983. In 1990, he became the reader (associate professor) in the
same department. He has been holding the post of Professor since 1998. He has taught courses on control systems and piecewise constant orthogonal functions, digital control systems, synchronous machine, transformer, Fourier transform, Laplace transform, and so on. His research interest includes automatic control in general and application of "alternative" orthogonal functions in power electronics and systems and control. He has published more than 65 research papers in different national and international journals and conferences. He is the principal author of the book *Triangular Orthogonal Functions for the Analysis of Continuous Time System* which was published twice—once by Elsevier, New Delhi, India, in 2007, and subsequently by Anthem Press, London, UK, in 2011.

Suchismita Ghosh, born in 1986, obtained her BTech (2008) from the Calcutta Institute of Engineering and Management, West Bengal University of Technology, India, and MTech (2010) from the Department of Applied Physics, University of Calcutta, India. She has been involved in research from her masters days (2009), and currently she is an assistant professor in the Department of Electrical Engineering, MCKV Institute of Engineering, West Bengal University of Technology, India. She has taught courses on
power electronics, basic electrical engineering, control systems, and electrical machines. Her research area includes automatic control in general and application of "alternative" orthogonal functions in systems and control. She is presently involved in research with Anish Deb and has published five research papers in international journals and national conferences.

1

Introduction

1.1 Evolution of Power Electronics

In the history of electrical technology spanning over the past four decades, a few developments have had a large impact upon the general area of power electronics (PE). These are the development and evolution of thyristors, microprocessors, microcontrollers, and so on. During the past few decades, we have seen the impact of PE devices, consisting of thyristors, gate turn-off thyristors (GTOs), power diodes, power transistors, power insulated gate field-effect transistors (IGFETs), power insulated gate bipolar transistors (IGBTs), metal–oxide–semiconductor field-effect transistors (MOSFETs), and so on in all aspects of modern motion control and power utilities [1]. The application of PE devices covers the low-power ranges such as in fractional horse power drives for robots as well as the high-power ranges involving Bay Area Rapid Transit (BART) trains of the United States or in Train á Grande Vitesse (TGV) trains of France. Today, complex control and decision strategies are implemented by PE devices with the help of microprocessors and microcontrollers to extract high performance and operational flexibility. In modern applications, useful features can be extracted even from a simple converter—inverter circuit, when operated as an uninterruptible power supply (UPS) or in high-voltage direct current (HVDC), where solid-state electronics are used in conjunction with PE devices [2,3].

Serious attention has been focused to evolve analytical techniques for predicting and explaining the performances of PE systems [4–6]. Many different techniques of varying degrees of complexities have been proposed to model a PE system. With the advancement in all disciplines during the past few decades, power electronics is now a major common factor in studies in control applications, electronics, and power utilities. In addition, the fields of instrumentation and computers are much involved and dependent on PE [7,8]. Hence, as shown in Figure 1.1, a star pattern has now emerged from a triangle pattern [7].

With the advancement of conventional silicon controlled rectifier (SCR), its GTO variety [9–11] was also manufactured in the 1960s. Its additional feature, that a negative pulse at the gate terminal turns the device off, has proven to be advantageous because it dispenses with the commutation circuitry of the conventional SCR when operated in a DC system. However, till

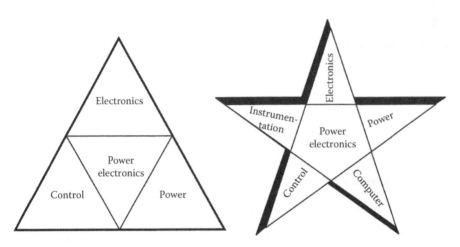

FIGURE 1.1
The changing pattern of PEs with regard to application areas.

the late 1970s, the power-handling capabilities of the GTOs were not much to credit their widespread application in the industry. However, during the last quarter of a century, technologists have been able to produce high-current GTOs suitable for critical applications involving high power.

Today, we have many types of PE devices satisfying the conflicting demands of various industries. The range of application of these devices is so wide as to make it unlikely that some new products will dominate all the existing ones. Certainly, no contender has yet met all the following specifications—and none possibly ever will, because some of the features are conflicting. These are:

- Zero leakage when "off"
- Zero impedance when "on"
- Zero switching time
- Zero switching losses, with no radiofrequency interference
- Very sensitive for ease of drive, but immune to spurious triggering
- Low cost
- Indefinite life: rugged
- Convenient packaging: isolated

1.2 Analysis of Power Electronic Circuits

The first mathematical analysis of the two-pulse power converter was carried out by Steinmetz in 1905. However, Dallenbach, who worked under Albert Einstein, was the first to propose the theoretical foundation of multipulse power converters. The theory and analysis of power converter circuits were enriched between 1930 and 1960 by the works of Muller-Lubeck, Marte and Winogard, Denontvignier, Schilling, Uhtmann, Kaganow, Wasserrals, and Rissik [12].

In 1940, Goodhue [13] presented an analysis technique using a rectifying mathematical operator. In his work, he applied his "rectifier calculus" to various rectifier circuits supplied with unbalanced sinusoidal/nonsinusoidal power line voltages. He represented the output voltage wave of a rectifier as a product of two voltage waveforms, one of which was a square wave.

With the exponential growth of the utility of PE systems, it became necessary to analyze and model such a system truthfully with the help of different available mathematical tools. However, each of the different analytical approaches has its own unique advantages as well as disadvantages.

A simple block diagram of a general PE converter is shown in Figure 1.2. The sole objective of the circuit in Figure 1.2 is to convert efficiently the input voltage or a set of voltages V_{IN}, having a particular waveform, to the output voltage or a set of voltages V_{OUT}, having a waveform different from V_{IN}. Normally, V_{IN} is a DC and single-phase or polyphase AC voltage vector. The output of the converter V_{OUT} may be a DC voltage of a different average value and a single-phase or polyphase AC voltage of a magnitude and frequency different from the input voltage.

The conversion in the PE converter is achieved by a set of semiconductor switches—usually thyristors, GTOs, or transistors—triggered in a specific sequence depending on the desired nature of the output voltage. Proper filters may be connected between the input and the output. The triggering or the firing sequence is controlled by the control signal c. But it may so happen that the nature of the switches and the connected load influence the conduction interval, and hence the operation becomes load dependent. This situation is indicated by the feedback signal f in Figure 1.2.

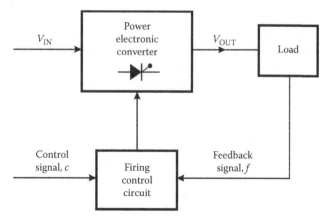

FIGURE 1.2
Block diagram of a general PE converter.

Due to the complexities mentioned above, the exact analysis of a PE system is seldom feasible, and the exact determination of all significant circuit variables as a function of time is difficult. Thus, the currently available analysis techniques usually look for approximate solutions that are obtained from a simplified linear or nonlinear representation of the system. These techniques are briefly stated in the following.

1.2.1 Fourier Series Technique

This is the oldest method employing orthogonal functions, namely, sine and cosine, which is useful in analyzing physical problems in various fields. In analyzing PE circuits, it is mostly used for phase control techniques [14].

If the load of the switching circuit is considered linear and the voltages applied to the load by the converter are known, the principle of superposition holds and Fourier series may be employed to determine the output variables successfully [15]. For each harmonic, the response is determined by solving the circuit for the respective harmonic equivalent load circuit. These individual responses are then summed to get the approximate output waveform.

However, it should be apparent that the analysis technique is really tedious and has to deal with many harmonic terms even in the case of a simple switching pattern of a PE converter. Most of the times, simplifications are made in the analysis, but this implies loss in accuracy.

1.2.2 Laplace Transform Method

This is one of the most popular as well as versatile methods for solving the problems of PE circuit analysis [16,17]. However, even in the straightforward application of the Laplace transform in most of the practical cases, the mathematical manipulations become quite involved.

In the application of the Laplace transform technique to chopper circuits, the output voltage of the PE controller, V_{OUT}, is represented as segments of a step function. When applied to phase-controlled circuits, V_{OUT} consists of segments of sine waveforms. The basic approach is to express V_{OUT} as a series, in which each term is the product of a sine wave and a switching function consisting of segments of a unit step function. Thus, each term represents a complete period or a quasi-period of the converter output voltage. Once the output voltage is determined, its Laplace transform is obtained by using the well-known formulas.

With V_{OUT} at our disposal, the output load current I_L can be expressed in terms of V_{OUT}, the load parameters, thyristor extinction angle, and so on in the transformed Laplace (s) domain. Then, applying the inverse Laplace transform yields an explicit function for the load current as an infinite series.

While the advantage of the method is its exactness, its disadvantages lie in the facts that the mathematics become too involved for determining the transforms and inverse transforms in the Laplace domain, even for simple circuits. In addition, the analysis has to be repeated if one is interested in finding out the output variables for the same load circuit but with different types of controller output voltages V_{OUT}.

1.2.3 Existence Function Technique

Existence functions formally and quantitatively describe the switching pattern of any PE converter. With existence functions, it is possible to obtain not only accurate mathematical formulations of the dependent quantities but also graphical constructions of dependent quantities and internal converter waveforms.

The existence function for a single switch, first introduced in the literature by Gyugyi and Pelly [18], assumes the unit value whenever the switch is closed, and it is zero whenever the switch is open. Within a specified time, the same switch may be closed or opened a number of times, having a variable repetition rate as well as "on"-time duration. If the switch has a fixed "on"-time and a fixed "off"-time duration, the switching pattern may be represented by the simplest unmodulated existence function. The more complex variety, which has different pulse durations and various interspersed zero times, is called a modulated existence function [19].

With the help of existence functions, even very complex switching patterns or switching matrices can be represented. The control variables of the converter system such as duty cycle, repetition frequency, and phase displacement are also taken into consideration. It was also found that the Fourier series technique was the best of all available methods to handle the spectral analysis of the existence functions [19]. Hence, all the inconveniences of the Fourier series technique are reflected in such an analysis. Further, while dealing with modulated existence functions, double and triple infinite series summations in the Fourier expansion result, aggravating the problem of analysis.

1.2.4 State Variable Method

This method is widely used, and its generality of system representation makes it applicable to a varied class of power converter circuits [20–22]. The basic principle of the state variable approach is that, given the differential equations describing a system and all the inputs to the system, there are a minimum number of variables called the state variables that completely determine the behavior of the system at all times. The state vector representation of a system expresses the system equations in vector form as a set of first-order differential equations [23].

The choice of variables for the state vector is not unique. However, in electrical systems, it is common practice to choose inductor currents and capacitor voltages as the state variables. The advantage of the state variable approach lies in the fact that it is systematic and thus convenient for computer simulation.

In circuits involving switches, the state variable approach is complicated due to the different modes caused by the switching of the devices. Thus, a simple circuit actually becomes very difficult to analyze because the analysis depends upon the conduction of a single device or a set of devices. The duration of the conduction periods of the semiconductor devices, which is not exactly known in most of the cases, adds to the complication of the analysis.

There are some variants of the general method of the state space approach, which deviate very little to warrant special attention [24].

1.2.5 Averaging Technique

This method is useful for analyzing low-frequency, small-signal responses of switching circuits. It is most suitable for circuits in which the power switching devices are operated in a short switching period compared to the response time of the output. Thus, it has been effectively applied to well-filtered, high-frequency DC-to-DC power converters [25,26].

Referring to Figure 1.2, the basic approach of this method is to average the inputs, V_{IN} and c, over a period of time that is short compared to the response time of the output V_{OUT}. It is assumed that the load does not affect device conduction so that the feedback signal f may be neglected. The system resulting from the averaging process may or may not be linear; based on this, the analysis proceeds by using standard applicable techniques such as Laplace transforms, state variables, and small-signal perturbation techniques for linearization.

1.2.6 z-Transform Analysis

Applying the z-transform or sampled-data technique to analyze PE converters is not uncommon [27–29]. In one approach [30], the semiconductor switch has been modeled as a sample-and-hold device. This model is then linearized to clearly show the time lag introduced into the system by the thyristor. In other approaches, the thyristor is modeled as a sampler of both the load current and the firing angle, and then a linearized discrete model is developed. The essential basic assumption of such treatments is that they allow limited excursion of the thyristor triggering angle.

The z-transform analysis [31,32], however, provides a solution that is discrete in nature. To obtain a continuous solution, one has to embark on the modified z-transform or submultiple sampling method [33,34]. Also, while handling truncated sinusoidal waveforms, flat-top pulse approximation affects the accuracy, with no appreciable reduction in mathematical exercise.

1.2.7 Other Methods of Analysis

In addition to the major techniques described later, there are several other ways that have been employed by some researchers at different times.

The switching function analysis [35] found its use because of the repetitive manner in which the devices in a PE converter are switched between conducting and blocking states. So, it seems to be reasonable that some sort of periodic function or rectangular pulses may simplify the converter representation. These functions, called switching functions, may switch between a finite number of states or may be multiplied by sinusoids to produce segments of sine waves. But switching functions, as with most analysis techniques, are not as easy to apply when the nature of the load current determines the conduction period through the feedback signal f as shown in Figure 1.2.

Considering its on–off properties, the thyristor could also be modeled as a binary device. Hence, application of Boolean algebra may find its use to simplify the waveform analysis or representation. This possibility has also been explored with some success [36,37]. These proposals are more systematic as well as organized representations of the switching function technique. When the thyristor is considered as a switch, the network topology for the system can also be drawn up, which will lead to a state space equation, the solution of which would give the performance of the system [38,39].

The bond graph, which was originally used to solve the problems of interconnected mechanical systems, is also applied to the analysis of PE systems. The switching of a PE device in a converter is graphed basically with a modulated transformer element, the "gain" of which is related as a function of the switching state [40].

In addition, describing function technique [41], the signal flow graph method [42,43] and symmetrical components approach [44] have also been applied for analyzing thyristor power converter circuits.

1.3 Search for a New Method of Analysis

It may be concluded that a tremendous amount of work has been carried out on the analysis [45] of static power converters. But, to date, no single technique has emerged as being clearly superior to all others. This fact has kept the search on for new methods to find out a uniquely advantageous general approach for analyzing the static power converter systems.

It is noted that switching functions, such as existence functions, can represent the output voltage of a static power converter in a simplified manner. But their ultimate dependence on Fourier analysis counterbalances any advantage gained by the simpler representation of the converter output voltage. However, the complexity of the Fourier series approach drives one to look for an alternative set of orthogonal functions, other than sine–cosine functions, which will be piecewise constant to suit our need. It could also be a worthwhile effort to search for a completely new method of analysis for PE systems using this alternative set of piecewise constant orthogonal functions. In what follows, we explore the suitability of Walsh functions [46], invented by J. L. Walsh way back in 1923, for the analysis of PE systems.

References

1. Rashid, M. H., *Power Electronics Handbook*, Academic Press, New York, 2001.
2. Hoft, R. G., The impact of microelectronics and microprocessors on power electronics and variable speed drives, IEE Conference Publication No. 234 "Power Electronics and Variable Speed Drives," London, May 1984.
3. Bose, B. K., *Microcomputer Control of Power Electronics and Drives*, IEEE Press, New York, 1987.
4. Hoft, R. G. and Casteel, J. B., Power electronic circuit analysis techniques, Proceedings of the 2nd IFAC Symposium on Control in Power Electronics and Electrical Drives, Düsseldorf, Germany, October 3–5, 1977, pp. 987–1024.
5. Shepherd, W. and Zhang, L., *Power Converter Circuits* (1st Ed.), Taylor & Francis, New York, 2010.
6. Rashid, M. H., *Power Electronics Circuits, Devices, and Applications* (3rd Ed.), Pearson Education, Delhi, 2004.
7. Datta, A. K., University–industry interaction in power electronics, Proceedings of the Conference on Power Electronics, IIT Kharagpur, West Bengal, May 1988.
8. Kloss, A., History of European power electronics, Proceedings of the 1st European Conference on Power Electronics and Applications, Brussels, Belgium, October 1985, pp. 1.1–1.6.
9. Van Ligten, R. H. and Navon, D., Base turn-off of p-n-p-n switches, IRE Wescon Convention Record, Part 3 on Electron Devices, Los Angeles, CA, August 1960, pp. 49–52.
10. Storm, H. F., Introduction to turn-off silicon controlled rectifiers, IEEE Conference Paper No. 63–321, Winter General Meeting, New York, January 27, 1963.
11. Wolley, E. D., Gate turn-off in p-n-p-n devices, *IEEE Trans. Electron Devices*, Vol. **ED-13**, No. 7, pp. 590–597, 1966.
12. Rissik, H., *Mercury-Arc Current Converters*, Sir Issac Pitman & Sons. Ltd., London, 1941.
13. Goodhue, W. M., The rectifier calculus, *AIEE Transactions*, Vol. **59**, pp. 687–691, 1940.
14. Shepherd, W., *Thyristor Control of AC Circuits*, Crosby Lockwood Staples, England, 1975.
15. Pelly, B. R., *Thyristor Phase-Controlled Converters and Cycloconverters*, Wiley Interscience, New York, 1971.
16. Fossard, A. J. and Clique, M., A general model for switched DC–DC converters including filters, Proceedings of the 2nd IFAC Symposium on Control in Power Electronics and Electrical Drives, Düsseldorf, Germany, October 3–5, 1977, pp. 117–126.
17. Rashid, M. H., Dynamic responses of DC chopper–controlled series motor, *IEEE Trans. Ind. Electron. Contr. Instrum.*, Vol. **IECI-28**, No. 4, pp. 323–330, 1981.

18. Gyugyi, L. and Pelly, B. R., *Static Power Frequency Changers*, Wiley Interscience, New York, 1976.
19. Wood, P., *Switching Power Converters*, Van Nostrand Reinhold Co., New York, 1981.
20. Ciccarella, G., de Carli, A. and la Cava, M., Analysis of SCR circuits via augmented state transition matrix, Proceedings of the 2nd IFAC Symposium on Control in Power Electronics and Electrical Drives, Düsseldorf, Germany, October 3–5, 1977, pp. 101–106.
21. Williams, B. W., Complete state-space digital computer simulation of chopper fed DC motor, *IEEE Trans. Ind. Electron. Contr. Instrum.*, Vol. **IECI-25**, No. 3, pp. 255–260, 1978.
22. El–Sátter, A. A. and Ahmed, F. Y., A separately excited DC motor supplied from a DC chopper supply, Proceedings of the 1st European Conference on Power Electronics and Applications, Brussels, Belgium, October 1985, pp. 3.289–3.293.
23. Ogata, K., *Modern Control Engineering* (5th Ed.), Prentice-Hall, New Delhi, 2010.
24. Nayak, P. H. and Hoft, R. G., Computer-aided steady-state analysis of thyristor DC drives on weak power systems, IEEE Conference Record of 11th Annual Meeting of Industry Application Society, Chicago, October 11–14, 1976, pp. 835–847.
25. Wester, G. W. and Middlebrook, R. D., Low-frequency characterisation of switched DC–DC converters, *IEEE Trans. Aero. Elect. Syst.*, Vol. **AES-9**, No. 3, pp. 376–385, 1973.
26. Middlebrook, R. D. and Cuk, S., A general unified approach to modelling switching converter power stages, Conference Record of IEEE Power Electronics Specialists Conference, Cleveland, OH, June 8–10, 1976, pp. 18–34.
27. Stojic, M. R., Microprocessor-based control of DC motors, in *Microprocessor-based Control Systems* (Ed. N. K. Sinha), D. Reidel Publishing Co., the Netherlands, 1986, pp. 131–155.
28. Casteel, J. B. and Hoft, R. G., Optimum PWM waveforms of a microprocessor controlled inverter, Proceedings of the IEEE Power Electronics Specialists Conference, Syracuse, June 13–15, 1978, pp. 243–250.
29. Ishida, K., Nakamura, K., Izmui, T. and Ohara, M., Microprocessor control of converter-fed DC motor drives, Proceedings of the IEEE/IAS Annual Meeting, San Francisco, CA, October 4–8, 1982, pp. 619–623.
30. Parrish, E. A. and McVey, E. S., A theoretical model for single phase silicon-controlled rectifier systems, *IEEE Trans. Automat. Contr.*, Vol. **AC-12**, No. 5, pp. 577–579, 1967.
31. Kuo, B. C., *Digital Control Systems* (2nd Ed.), Oxford University Press, New York, 1992.
32. Ogata, K., *Discrete-Time Control Systems* (2nd Ed.), Pearson Education, Delhi, 2001.
33. Bühler, H., Study of a DC chopper as a sampled system, Proceedings of the 2nd IFAC Symposium on Control in Power Electronics and Electrical Drives, Düsseldorf, Germany, October 3–5, 1977, pp. 67–77.
34. Mochizuki, T., Tanaka, Y. and Hyodo, M., A simple method for a quick-response chopper PWM system with small steady-state error, Proceedings of the 2nd IFAC Symposium on Control in Power Electronics and Electrical Drives, Düsseldorf, Germany, October 3–5, 1977, pp. 79–86.

35. Novotny, D. W., Switching function representation of polyphase inverters, Conference Record on 10th Annual Meeting of Industry Applications Society, Atlanta, GA, 1975, pp. 823–831.
36. Deb, A. and Datta, A. K., A sequential machine model of a thyristor, *Int. J. Electron.*, Vol. **58**, No. 1, pp. 151–157, 1985.
37. Ho, H. H., Improved logic model of thyristor, *Proc. IEE*, Vol. **121**, No. 5, pp. 345–347, 1974.
38. Revankar, G. N., Topological approach to thyristor-circuit analysis, *Proc. IEE*, Vol. **120**, No. 11, pp. 1403–1405, 1973.
39. Revankar, G. N. and Mahajan, S. A., Digital simulation for mode identification in thyristor circuits, *Proc. IEE*, Vol. **120**, No. 2, pp. 269–272, 1973.
40. Masada, E., Hiraiwa, Y., Hayafune, K. and Tamura, M., The bond graph model of electronic power converter, Proceedings of the 1st European Conference on Power Electronics and Applications, Brussels, Belgium, October 1985, pp. 1.103–1.107.
41. Haneda, H. and Marunashi, T., DC and AC analysis of thyristor circuits by coordinate transformation and describing-function method, Proceedings of the 2nd IFAC Symposium on Control in Power Electronics and Electrical Drives, Düsseldorf, Germany, October 3–5, 1977, pp. 127–134.
42. Subrahmanyam, V., Subbarayudu, D. and Rao, M. V. C., On the utility of signal flow graphs in the analysis of current controlled induction motor, Proceedings of the 2nd IFAC Symposium on Control in Power Electronics and Electrical Drives, Düsseldorf, Germany, October 3–5, 1977, pp. 455–462.
43. Subrahmanyam, V., State space approach to performance analysis of separately excited DC motor using state transition signal flow graph technique, *J. Inst. Eng.*, Vol. **67**, No. Pt EL-1, pp. 21–24, 1986.
44. Bausz, I., Analysis of forced commutation processes of a bridge inverter by means of symmetrical components, Proceedings of the 2nd IFAC Symposium on Control in Power Electronics and Electrical Drives, Düsseldorf, Germany, October 3–5, 1977, pp. 37–44.
45. Deb, A. and Datta, A. K., On analytical techniques of power electronic circuit analysis, *IETE Tech. Rev.*, Vol. **7**, No. 1, pp. 25–32, 1990.
46. Walsh, J. L., A closed set of normal orthogonal functions, *Am. J. Math.*, Vol. **45**, pp. 5–24, 1923.

2

An Alternative Class
of Orthogonal Functions

Orthogonal properties [1] of the familiar sine–cosine functions have been known to the academic community for more than two centuries. But using such functions in an elegant manner to solve complex analytical problems was initiated by the work of the famous mathematician Baron Jean Baptiste Joseph Fourier [2]. With the passage of time, Fourier analysis established itself as one of the major analytical techniques.

In many areas of electrical engineering, the basis for any analysis is a system of sine–cosine functions. This is mainly due to the desirable properties of frequency domain representation of a large class of functions encountered in theoretical and practical aspects of engineering design.

In the fields of circuit analysis, control theory, communication, and analysis of stochastic problems, examples are found extensively where the orthogonal properties of such a "complete" system lend themselves to particularly attractive solutions. But with the application of digital techniques and semiconductor technology in these areas, awareness for other more general complete systems of orthogonal functions has been developed. This "new" class of functions, though not possessing some of the desirable properties of sine–cosine functions in linear time-invariant networks, has other advantages in the context of digital technology. This class of orthogonal functions is piecewise constant binary valued, thus resembling the high–low switching characteristics of semiconductor devices, and therefore seemed to be suitable in analyzing systems using two-state digital logic.

2.1 Orthogonal Functions and Their Properties

Any time function can be synthesized completely to a tolerable degree of accuracy by using a set of orthogonal functions. For such accurate representation of a time function, the orthogonal set should be "complete" [1,3,4].

Let a time function $f(t)$, defined over a time interval $[0, T)$, be represented by an orthogonal function set $\mathbf{G}_{(n)}(t)$. Then

$$f(t) \approx \sum_{n=0}^{m-1} c_n g_n(t) = c_0 g_0 + c_1 g_1 + c_2 g_2 + \cdots + c_{m-1} g_{m-1} \qquad (2.1)$$

where:
   m is a very large number
   c_n is the coefficient or weight connected to the $(n + 1)$th member of the complete orthogonal set $\mathbf{G}_{(n)}(t)$

The member of the function set $\mathbf{G}_{(n)}(t)$ is said to be orthogonal in the interval $0 \le t \le T$ if for any positive integral values of m and n, we have

$$\int_0^T g_m(t) g_n(t) dt = \delta_{mn} \qquad (2.2)$$

where:
   δ_{mn} is the Kronecker delta
   $\delta_{mn} = 0$ for $m \ne n$ and δ_{mn} is the constant for $m = n$

When $\delta_{mn} = 1$, the set is said to be an orthonormal set. Incidentally, any orthogonal set may easily be normalized to convert it to an orthonormal set.

Since only a finite number of terms of the series $\mathbf{G}_{(n)}(t)$ can be considered for practical realization of any time function $f(t)$, the right-hand side (RHS) of Equation 2.1 has to be truncated so that we have

$$f(t) \approx \sum_{n=0}^{m-1} c_n g_n(t) \qquad (2.3)$$

When m is large, the accuracy of representation is good enough for all practical purposes. In addition, it is necessary to choose the coefficients c_n in such a manner that the mean integral square error (MISE) [3] for such approximation is guaranteed to be minimum. Thus,

$$\text{MISE} = \frac{1}{T} \int_0^T \left[f(t) - \sum_{n=0}^{m-1} c_n g_n(t) \right]^2 dt \qquad (2.4)$$

This is realized by making

$$c_n = \frac{1}{\delta_{mn}} \int_0^T f(t) g_n(t) dt \qquad (2.5)$$

2.2 Haar Functions

Historically, the Hungarian mathematician Alfred Haar [5] was the first to propose in 1910 a complete set of piecewise constant binary-valued orthogonal functions χ shown in Figure 2.1. Haar's set [6] is such that the formal expansion of a given continuous function in terms of these new functions converges uniformly to the given function [7].

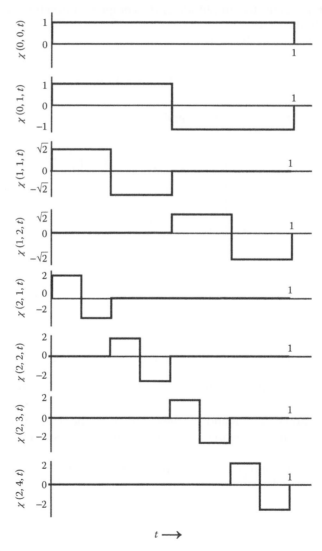

FIGURE 2.1
A set of Haar functions.

The piecewise constant nature of this "new" class of bivalued functions attracted many researchers to explore its appropriate application in the areas of power electronic (PE), digital signal processing, control theory, communication, and many more.

The binary-valued Haar function set is not only the pioneer of the piecewise constant basis function (PCBF) family but also the simplest wavelet [8,9] showing the required scaling and shifting property along with its orthogonal nature.

Haar functions have three possible states 0, $+A$, and $-A$, where A is a function of $\sqrt{2}$. Thus, the amplitude of the component functions varies with their place in the series.

The first member of the set is

$$\chi(0,0,t) = 1, \qquad t \in [0, 1) \tag{2.6}$$

while the general term for other members is given by

$$\chi(j,n,t) = \begin{cases} 2^{\frac{j}{2}}, & \dfrac{(n-1)}{2^j} \le t < \dfrac{\left(n - \frac{1}{2}\right)}{2^j} \\[3mm] -2^{\frac{j}{2}}, & \dfrac{\left(n - \frac{1}{2}\right)}{2^j} \le t < \dfrac{n}{2^j} \\[3mm] 0, & \text{elsewhere} \end{cases} \tag{2.7}$$

where:
$j, n,$ and m are integers governed by the relations $0 \le j \le \log_2(m)$ and $1 \le n \le 2^j$

The number of members in the set is of the form $m = 2^p$, p being a positive integer.

Following Equations 2.6 and 2.7, the members of the set of Haar functions can be obtained in a sequential manner. In Figure 2.1, p is taken to be 3, thus giving $m = 8$.

Let a time function $f(t)$, defined over a time interval $[0, T)$, be represented by an orthogonal Haar function set $\chi_{(n)}(t)$. Then, according to Equation 2.1,

$$f(t) = \sum_{n=0}^{\infty} c_n \chi_n(t) = c_0 \chi_0 + c_1 \chi_1 + c_2 \chi_2 + \cdots + c_n \chi_n + \cdots$$

2.3 Rademacher and Walsh Functions

Inspired by Haar, the search for similar orthogonal functions became important to some mathematicians, and in 1922 the German mathematician H. Rademacher presented another set of two-valued orthogonal functions [10] shown in Figure 2.2. Nearly at the same time, the American mathematician J. L. Walsh independently proposed yet another binary-valued complete set of normal orthogonal functions Φ [11], later named Walsh functions.

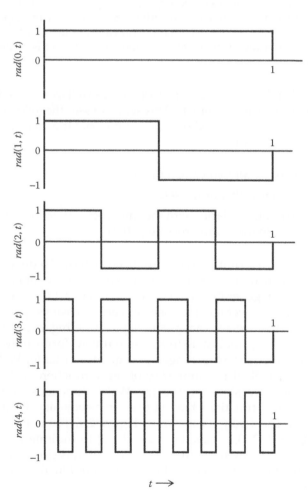

FIGURE 2.2
A set of Rademacher functions.

In his original paper, Walsh pointed out that "... Haar's set is, however, merely one of an infinity of sets which can be constructed of functions of this same character." While proposing his new set of orthonormal functions Φ, Walsh wrote, "... each function ϕ takes only the values +1 and −1, except at a finite number of points of discontinuity, where it takes the value zero."

> "The chief interest of the set Φ lies in its similarity to the usual (e.g. sine, cosine, Strum–Liouville, Legendre) sets of orthogonal functions, while the chief interest of the set χ lies in its dissimilarity to these ordinary sets. The set Φ shares with the familiar sets the following properties, none of which is possessed by the set χ: the n-th function has $n - 1$ zeroes (or better, sign-changes) interior to the interval considered, each function is either odd or even with respect to the mid-point of the interval, no function vanishes identically on any sub-interval of the original interval, and the entire set is uniformly bounded."

It was found that the Rademacher function set as well as a true subset of the Walsh function set was incomplete. This set possesses the following properties, all of which are not shared by other orthogonal functions belonging to the same class. These are:

1. Its members are all two-valued functions.
2. It is a complete orthonormal set.
3. It has a striking similarity with the sine–cosine functions, primarily with respect to their zero-crossing patterns.

In the late 1960s, Harmuth [12] introduced a term "sequency," which characterizes Walsh and other nonsinusoidal orthogonal functions. Sequency (z) is a generalized term for frequency (f) and is defined as one-half of the average number of zero crossings per unit time interval. From this, we see that frequency can be regarded as a special measure of sequency applicable to sinusoidal waveforms only. In Figure 2.3, the Walsh functions are arranged in sequency order (or Walsh order). To have a better look at the similarity of these functions with the sine–cosine functions, we superimpose the Walsh functions of Figure 2.3 with sine or cosine functions having the same sequency–frequency correspondence, respectively, as shown in Figure 2.4. The notations $sal\ (i, t)$ and $cal\ (i, t)$, $i = 1, 2, 3,...$ have been used [12] to indicate to the sine or cosine function that any particular Walsh function is closely related. The first Walsh function of the set has been named $wal\ (0, t)$, which is a unit step function.

In addition to the sequency ordering, there are two more ordering conventions that are in common use. While the sequency order shown in Figure 2.3

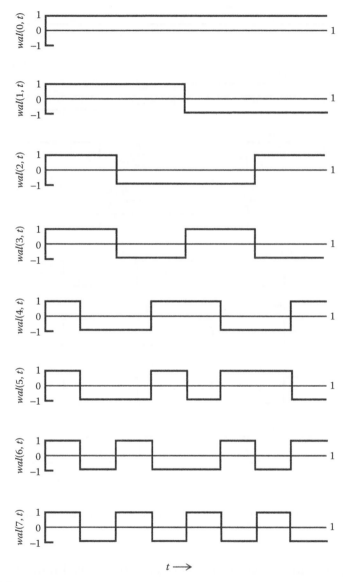

FIGURE 2.3
A set of Walsh functions arranged in an order proposed by J. L. Walsh.

is the closest to our practical experience with other orthogonal functions such as sinusoidal functions, it does not offer the most convenient order for analytical and computational purposes. That is why the dyadic order, first used by Paley [13], and the natural order, first proposed by Henderson [14], became popular.

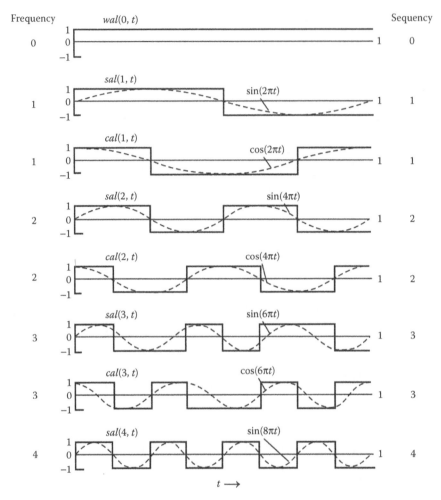

FIGURE 2.4
Walsh functions and sine–cosine functions having the same sequency–frequency correspondence respectively.

The dyadic order is obtained by the generation of Walsh functions from successive Rademacher functions. Since it offers advantages for analytical and computational purposes, this ordering was chosen for possible use in analyzing power converter circuits. The first 16 Walsh functions, arranged in dyadic order, are shown in Figure 2.5.

It is interesting to note that some of the patterns of individual Walsh functions are quite familiar because they appear in several ancient motifs and designs. In ancient temples, palaces, and gothic structures, such square-wave patterns are not uncommon. Chessboard or checkerboard designs are nothing but two-dimensional Walsh functions, whereas the

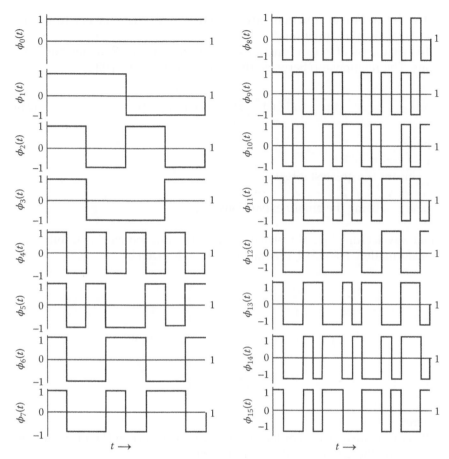

FIGURE 2.5
First 16 Walsh functions arranged in dyadic order. (Reprinted from Chen, C. F. and Hsiao, C. H., *Int. J. Syst. Sci.*, 6, 833–858, 1975. With permission.)

famous Rubik's Cube [15] is a perfect example of a three-dimensional Walsh function.

As indicated by Harmuth [12], the complete set of Walsh functions was used by John A. Barrett at around 1900 AD for the transposition of conductors in open wire lines. Until then, Rademacher functions were used for this purpose. The transposition of conductors according to Barrett's scheme was almost a standard practice in 1923, when J. L. Walsh introduced them formally in a well-organized manner into the literature [11].

As stated by Walsh, there are many possible orthogonal function sets of this kind, and several researchers, in later years, have suggested orthogonal series [16,17] formed with the help of combinations of the well-known sequency functions.

The set of Walsh functions are related to both Rademacher functions and Haar wavelet functions.

In fact, the set of Walsh functions can be derived from Rademacher functions. The relationship between the two sets is given by [18]

$$\phi_n(t) = \left[rad_q(t) \right]^\alpha \left[rad_{q-1}(t) \right]^\beta \left[rad_{q-2}(t) \right]^\gamma \cdots \qquad (2.8)$$

where:

$q = \left[\log_2 n \right] + 1$

$[.]$ means the greatest integer

$n = \alpha \, 2^{q-1} + \beta \, 2^{q-2} + \gamma \, 2^{q-3} + \cdots$

$\alpha, \beta, \gamma, \ldots$ being the binary expansions of n.

Haar wavelet functions are related to the Walsh function set via the following transformation. For $m = 4$, the transformation is given by

$$\chi_{(4)} = \begin{bmatrix} 1 & 0 & 0 & 0 \\ 0 & 1 & 0 & 0 \\ 0 & 0 & \dfrac{1}{\sqrt{2}} & \dfrac{1}{\sqrt{2}} \\ 0 & 0 & \dfrac{1}{\sqrt{2}} & -\dfrac{1}{\sqrt{2}} \end{bmatrix} \Phi_{(4)} \triangleq \mathbf{P}_{(4)} \Phi_{(4)} \qquad (2.9)$$

where:

$\mathbf{P}_{(4)}$ is a square matrix of order 4 relating the orthonormal Haar function set and the orthonormal Walsh function set.

P has a property such that

$$\mathbf{P}_{(4)}^\mathrm{T} \mathbf{P}_{(4)} = \mathbf{P}_{(4)}^2 = \begin{bmatrix} 1 & 0 & 0 & 0 \\ 0 & 1 & 0 & 0 \\ 0 & 0 & 1 & 0 \\ 0 & 0 & 0 & 1 \end{bmatrix}$$

which shows **P** is an orthogonal matrix.

In general,

$$\mathbf{P}_{(m)}^\mathrm{T} \mathbf{P}_{(m)} = \mathbf{P}_{(m)}^2 = \mathbf{I}_{(m)} \qquad (2.10)$$

for any values of $m = 2^p$, where $p = 1, 2, 3, \ldots$.

2.3.1 Representation of a Function Using Walsh Functions

Example 2.1

Consider the function $f_1(t) = \sin(\pi t)$ (Program 1 in Appendix B). For $m = 8$ and $T = 1$ second, using Equation 2.5, we can compute the Walsh coefficients c_i's to approximate the function $f_1(t)$ in Walsh domain. Thus,

$$f_1(t) \approx [0.6366 \; 0 \; 0 \; -0.2637 \; 0 \; -0.1266 \; -0.0525 \; 0] \Phi_{(8)}(t) \triangleq C \, \Phi_{(8)}(t) \quad (2.11)$$

From the above equation, we plot the Walsh expansion of $f_1(t)$ along with the exact curve shown in Figure 2.6.

Example 2.2

Consider the function $f_2(t) = \exp(-t)$ (Program 2 in Appendix B). For $m = 8$ and $T = 1$ second, using Equation 2.5, we can compute the Walsh coefficients d_i's to approximate the function $f_2(t)$ in Walsh domain. Thus,

$$f_2(t) \approx [0.6321 \; 0.1548 \; 0.0786 \; 0.0193 \; 0.0395 \; 0.0097 \; 0.0049 \; 0.0012] \Phi_{(8)}(t) \quad (2.12)$$
$$\triangleq D \, \Phi_{(8)}(t)$$

Using the above equation, we plot the Walsh expansion of $f_2(t)$ along with the exact curve shown in Figure 2.7.

Since the approximation in Walsh domain always gives a staircase result, it is apparent that an increase in m will reduce the approximation error.

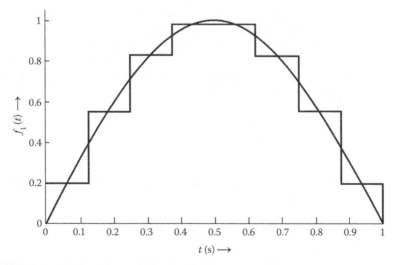

FIGURE 2.6
Comparison between the exact function $\sin(\pi t)$ and its Walsh approximation.

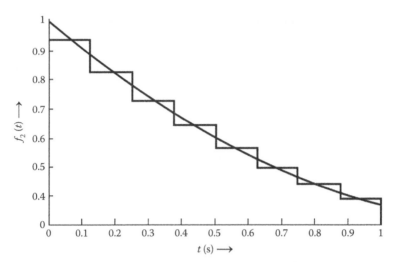

FIGURE 2.7
Comparison between the function exp(−t) and its Walsh approximation.

2.4 Block Pulse Functions and Their Applications

During the nineteenth century, the most important functions used for communication were block pulses [12]. Voltage and current pulses, such as Morse code signals, were generated by mechanical switches, amplified by relays and detected by various magnetomechanical devices. However, the set of block pulses received less attention from the mathematicians as well as from the application engineers possibly due to their apparent incompleteness until recently. But disjoint and orthogonal properties of such a function set were well known. A set of block pulse functions (BPFs) is shown in Figure 2.8.

The $(i + 1)$th member of an m-set BPF, $\psi_i(t)$ $(i = 0, 1, 2, 3,..., m − 1)$, is defined in the semi-open interval $t \in [0, T)$ as

$$\psi_i(t) = \begin{cases} 1, & \text{for } iT/m \le t < (i + 1)T/m \\ 0, & \text{otherwise} \end{cases} \quad (2.13)$$

where:
  m is an arbitrary positive integer and the duration of each block pulse is $T/m = h$ seconds

For evaluation of BPF expansion coefficients of a function using Equation 2.5, the Kronecker delta $\delta_{mn} = h$.

The BPF set is a complete orthogonal function set [4]. Though it is not an orthonormal set, it can easily be normalized by defining the component functions in the interval $[0, T)$ as

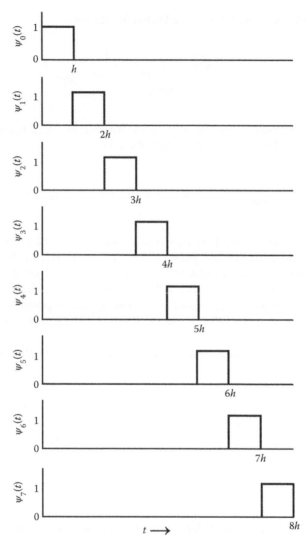

FIGURE 2.8
A set of eight BPFs.

$$\psi_{i\text{N}}(t) = \begin{cases} \dfrac{1}{\sqrt{h}}, & \text{for } ih \le t < (i+1)h \\ 0, & \text{otherwise} \end{cases}$$

That is,

$$\psi_{i\text{N}}(t) = \frac{1}{\sqrt{h}}\psi_i(t)$$

If T is taken to be 1 second, like the Walsh function set,

$$\psi_{iN}(t) = \begin{cases} \sqrt{m}, & \text{for } \dfrac{i}{m} \leq t < \dfrac{i+1}{m} \\ 0, & \text{otherwise} \end{cases}$$

Let a time function $f(t)$, defined over a time interval $[0, T)$, be represented by a BPF set $\Psi_{(n)}(t)$. Then, according to Equation 2.1,

$$f(t) = \sum_{n=0}^{\infty} c_n \psi_n(t) = c_0 \psi_0 + c_1 \psi_1 + c_2 \psi_2 + \cdots + c_n \psi_n + \cdots$$

A one-to-one relationship between Walsh function and BPF, obvious from Figures 2.5 and 2.8, was first presented by Wu et al. [19] in 1976 and was published later in a journal [20]. In their work, Chen et al. used block pulse operational matrices for simplifying their approach in Walsh domain analysis, but never embarked on an all-block pulse approach possibly because of their contention that the BPFs, at least in theory, did not form a complete set until then.

Haar wavelet functions are related to the BPF set via the following relation for $m = 4$.

$$\chi_{(4)} = \begin{bmatrix} 1 & 1 & 1 & 1 \\ 1 & 1 & -1 & -1 \\ \sqrt{2} & -\sqrt{2} & 0 & 0 \\ 0 & 0 & \sqrt{2} & -\sqrt{2} \end{bmatrix} \Psi_{(4)} \triangleq Q_{(4)} \Psi_{(4)} \tag{2.14}$$

where:
$Q_{(4)}$ is the Haar matrix of order 4 and it has a property such that

$$Q_{(4)}^T Q_{(4)} = 4 I_{(4)}$$

So, in general,

$$Q_{(m)}^T Q_{(m)} = m I_{(m)} \tag{2.15}$$

Had we used a normalized BPF set, instead of the traditional BPF set in Equation 2.14, the result of Equation 2.15 would have been a unit matrix of order m, making the relational matrix $Q_{(m)}$ orthogonal.

The relationship between a Walsh function set $\Phi_{(m)}$ of order m and an m-set BPF is

$$\Phi_{(m)} = \mathbf{W}_{(m)}\Psi_{(m)} \tag{2.16}$$

where:

$\mathbf{W}_{(m)}$ is an $m \times m$ square matrix known as the Walsh matrix. The Walsh matrix of order 4 is

$$\mathbf{W}_{(4)} = \begin{bmatrix} 1 & 1 & 1 & 1 \\ 1 & 1 & -1 & -1 \\ 1 & -1 & 1 & -1 \\ 1 & -1 & -1 & 1 \end{bmatrix} \tag{2.17}$$

and it has a property such that

$$\left. \begin{aligned} \mathbf{W}_{(m)}^{\mathrm{T}} &= \mathbf{W}_{(m)} \\ \text{and} \\ \mathbf{W}_{(m)}^{2} &= m\,\mathbf{I}_{(m)} \end{aligned} \right\} \tag{2.18}$$

where:

$\mathbf{I}_{(m)}$ is the identity matrix of order m

Here also, had we used the normalized BPF set, the result of Equation 2.18 would have been a unit matrix only. Hence, in Equation 2.16, the relational square matrix connecting the Walsh function set to the normalized BPF set would have been orthogonal. Haar function set, BPF set, and Walsh functions are related to one another via similarity transformation.

2.4.1 Representation of a Function as a Linear Combination of BPFs

Example 2.3

Consider the function $f_1(t) = \sin(\pi t)$ (Program 1 in Appendix B). For $m = 8$ and $T = 1$ second, using Equation 2.5, we can compute the BPF coefficients c_i''s to represent the function $f_1(t)$ in BPF domain as

$$f_1(t) \approx [0.1938\ 0.5520\ 0.8261\ 0.9745\ 0.9745\ 0.8261\ 0.5520\ 0.1938]\Psi_{(8)}(t)$$
$$\triangleq \mathbf{C}'\Psi_{(8)}(t) \tag{2.19}$$

From the above equation, we can plot the BPF expansion of $f_1(t)$ along with the exact curve shown in Figure 2.6 for comparison. A point to be noted is that since Walsh function and BPF sets [20,21] are related via similarity transformation, for the same m and T, a solution of any problem using any of these two sets always produces the same result.

Example 2.4

Consider the function $f_2(t) = \exp(-t)$ (Program 2 in Appendix B). For $m = 8$ and $T = 1$ second, using Equation 2.5, we can determine the BPF coefficients d_i''s to approximate the function $f_2(t)$ as

$$f_2(t) \approx [0.9400\ 0.8296\ 0.7321\ 0.6461\ 0.5702\ 0.5032\ 0.4440\ 0.3919]\Psi_{(8)}(t)$$

$$\triangleq \mathbf{D}'\ \Psi_{(8)}(t)$$

(2.20)

From the above equation, we can plot the BPF expansion of $f_2(t)$ along with the exact curve shown in Figure 2.7 for comparison.

Since Walsh and BPF analyses always give the same result, any analysis based on these two function sets yields the same MISE. Only the application suitability justifies the choice for selecting any of these two sets.

Like the Walsh function set, the BPF set is also complete. Its completeness and convergent properties were investigated by Kwong and Chen [4]. Surprisingly, it was found that not only the BPF set was complete, but it was more fundamental, generalized, and efficient than other orthogonal function sets of the same class. Several fascinating properties of the BPFs are found by the researchers [22–27], which are as follows:

1. It forms a complete orthogonal set that could easily be normalized.
2. It is computationally much simpler, and yet allows the same numerical accuracy as that obtained by Walsh function approach.
3. Any number of BPFs can be used to form a complete set, while Walsh functions have to satisfy the constraint $m = 2^p$ ($p = 1, 2, 3,...$), the number of component functions, to form a set suitable for analytical manipulations.
4. It needs less computation time as well as computer memory space in comparison with Walsh functions.
5. It is related to Haar, Rademacher, Walsh, and other similar functions by linear transformation matrices [28]. All piecewise constant orthogonal functions can be expressed as a linear combination of BPFs, but the reverse is not true in general. This makes the BPF set the most fundamental of all orthogonal functions of its class.

Now we compare the two function sets, namely, Walsh function and BPF. Table 2.1 presents their qualitative comparison.

With these salient features well identified, the so-called 'incomplete' set of BPFs was used by research workers with greater frequency in

TABLE 2.1

Basic Properties

Properties	BPFs	Walsh Functions
Piecewise constant	Yes	Yes
Orthogonal	Yes	Yes
Finite	Yes	Yes
Disjoint	Yes	No
Orthonormal	No, but can easily be normalized	Yes
Implementation	Easily implementable	Not easily implementable
Coefficient determination of $f(t)$	Involves integration of $f(t)$ and scaling	Involves only integration of $f(t)$
Accuracy of analysis	Provides staircase solution; accuracy same as Walsh analysis	Provides staircase solution; accuracy same as BPF analysis

different application areas [29]: analysis and synthesis of dynamic systems [23,25], bilinear system identification [26], fractional and operational calculus [27], multipoint boundary value problems [30], analysis of scaled systems [31], and solution of integral equations [32], just to name a few. Indeed, there are a score of other areas where this orthogonal set is finding its deserved use [33–35].

2.5 Slant Functions

A special orthogonal function series known as the slant function series was introduced by Enomoto and Shibata [36] for image transmission analysis. These functions have a finite but a large number of possible states as shown in Figure 2.9. These functions are similar to the general form of the sequency ordering of Walsh functions. The superiority of this function set lies in its transform characteristics that permit a compaction of the image energy to only a few transformed samples. Thus, the efficiency of image data transmission in this form is improved. As expected, the slant functions are related to the Walsh functions through the algebra of transformation matrices.

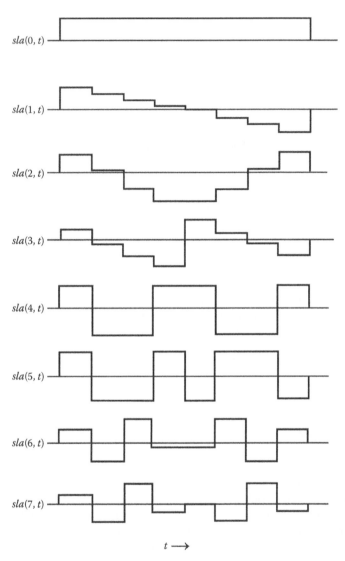

FIGURE 2.9
A set of slant functions.

2.6 Delayed Unit Step Functions

Delayed unit step functions (DUSFs), shown in Figure 2.10, were suggested
by Hwang [37] in 1983. Though it is not of much use due to its dependency
on BPFs, shown by Deb et al. [22], it certainly deserves to be included in the
record of PCBFs as a new variant.

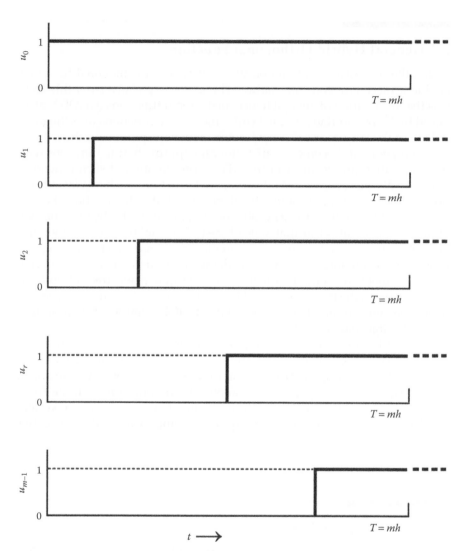

FIGURE 2.10
A set of DUSF for m-component functions.

$$u_i(t) = \begin{cases} 1 & t \geq ih \\ 0 & t < ih \end{cases}$$

where:
$i = 0, 1, 2, \ldots, (m-1)$

2.7 General Hybrid Orthogonal Functions

So far, the discussion centered on different types of orthogonal functions having a piecewise constant nature. The major departure from this class was the formulation of general hybrid orthogonal functions (GHOFs) introduced by Patra and Rao [38–40]. While sine–cosine functions or orthogonal polynomials can represent a continuous function quite well, these functions/polynomials become unsatisfactory for approximating functions with discontinuities, jumps, or dead time. For representation of such functions, undoubtedly piecewise constant orthogonal functions such as Walsh function or BPF can be used more advantageously. But with functions having both continuous nature and a number of discontinuities in the time interval of interest, it is quite clear that none of the orthogonal functions/polynomials is of continuous nature, or, for that matter, piecewise constant orthogonal functions are suitable if a reasonable degree of accuracy is to be achieved. Hence, to meet the combined features of continuity and discontinuity in such situations, the framework of GHOF proposed by Patra and Rao seemed to be more appropriate. Thus, the system of GHOF forms a hybrid basis that is both flexible and general.

Patra and Rao successfully applied GHOFs to analyze linear time-invariant systems, converter-fed and chopper-fed DC motor drives, and the area of self-tuning control. The main disadvantage of GHOF seems to be the required *a priori* knowledge about the discontinuities in the function, which are to be matched with the segment boundaries of the system of GHOF to be chosen. This also requires a complex algorithm for better results.

2.8 Sample-and-Hold Functions

In 1995, a pulse-width modulated version of the BPF set was presented by Deb et al. [41,42], where the pulse width of the component functions of the BPF set was gradually increased (or decreased) depending on the nature of the square-integrable function to be handled.

In 1998, a further variant of the BPF set was proposed by Deb et al. [43], which was called sample-and-hold function (SHF) set, and the same was utilized for the analysis of sampled data systems with zero-order hold (ZOH). The set of SHFs approximates any square-integrable function in a piecewise constant manner, and it proved to be more convenient for solving problems associated with sample-and-hold control systems.

Any square-integrable function $f(t)$ may be represented by an SHF set in the semi-open interval $[0, T)$ by considering

$$f_i(t) \approx f(ih), \qquad i = 0, 1, 2, \ldots, (m-1) \tag{2.21}$$

where:

h is the sampling period $(= T/m)$

$f_i(t)$ is the amplitude of the function $f(t)$ at time ih

$f(ih)$ is the first term of the Taylor series expansion of the function $f(t)$ around the point $t = ih$, because for a ZOH, the amplitude of the function $f(t)$ at $t = ih$ is held constant for the duration h

SHFs are similar to BPFs in many aspects. In fact, if we call the conventional BPF set an optimal BPF set with regard to minimized MISE for function approximation, the set of SHF may be regarded as a nonoptimal BPF set.

The $(i + 1)$th member of a set of SHFs, $\mathbf{S}_{(m)}(t)$, composed of m component functions, is defined as

$$s_i(t) = \begin{cases} 1, & \text{for } ih \le t < (i+1)h \\ 0, & \text{elsewhere} \end{cases} \tag{2.22}$$

where:

$i = 0, 1, 2, \ldots, (m-1)$

Considering the nature of the SHF set, which is a look-alike of the BPF set (Figure 2.8), it is easy to conclude that this set is orthogonal as well as complete in $t \in [0, T)$ like the BPF set. However, the special property of the SHF is revealed only by using the sample-and-hold concept in the derivation of the required operational matrices. If a time signal $f(t)$ is fed to a sample-and-hold device as shown in Figure 2.11, the output of the device approximates $f(t)$ as per Equation 2.21.

A square-integrable time function $f(t)$ may be expanded into an m-term SHF series in $t \in [0, T)$ as

$$f(t) \approx \sum_{i=0}^{m-1} f_i s_i(t) = (f_0 \ f_1 \ f_2 \ \ldots \ f_i \ \ldots f_{m-1}) \mathbf{S}_{(m)}(t) \qquad \text{for } i = 0, 1, 2, \ldots, (m-1) \tag{2.23}$$

$$\underline{\Delta} \ \mathbf{F}_{(m)}^{T} \mathbf{S}_{(m)}(t)$$

where:

$f_i = f(ih)$, the $(i + 1)$th sample of the function $f(t)$

$\mathbf{F}_{(m)}$ is considered to be a column vector

In fact, f_i's are the samples of the function $f(t)$ with the sampling period h.

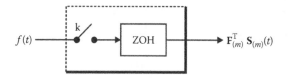

FIGURE 2.11
A sample-and-hold device.

2.9 Triangular Functions

In 2003, orthogonal triangular functions (TFs) were introduced by Deb et al. [44], and the same were applied to control system-related problems including analysis and system identification.

The set of TFs approximates any square-integrable function in a piecewise linear manner. From this basic property, relevant theories were developed [45,46] to make this set powerful enough to deal with control system-related problems [44–48].

From a set of BPFs, $\Psi_{(m)}(t)$, we can generate two sets of orthogonal TFs, namely, $\mathbf{T1}_{(m)}(t)$ and $\mathbf{T2}_{(m)}(t)$, such that

$$\Psi_{(m)}(t) = \mathbf{T1}_{(m)}(t) + \mathbf{T2}_{(m)}(t) \tag{2.24}$$

which is shown in Figure 2.12.

Figure 2.13 shows the TF sets, $\mathbf{T1}_{(m)}(t)$ and $\mathbf{T2}_{(m)}(t)$, where m has been chosen arbitrarily as 8. These two TF sets are complementary to each other. For convenience, we call $\mathbf{T1}_{(m)}(t)$ the left-handed triangular function (LHTF) vector and $\mathbf{T2}_{(m)}(t)$ the right-handed triangular function (RHTF) vector.

Using the component functions, we could express the m-set TF vectors as

$$\mathbf{T1}_{(m)}(t) \underline{\Delta} \left[T1_0(t)\ T1_1(t)\ T1_2(t) \ldots T1_i(t) \ldots T1_{(m-1)}(t) \right]^T$$

$$\mathbf{T2}_{(m)}(t) \underline{\Delta} \left[T2_0(t)\ T2_1(t)\ T2_2(t) \ldots T2_i(t) \ldots T2_{(m-1)}(t) \right]^T$$

The $(i + 1)$th component of the LHTF vector $\mathbf{T1}_{(m)}(t)$ is defined as

$$T1_i(t) = \begin{cases} 1 - (t - ih)/h, & \text{for } ih \leq t < (i+1)h \\ 0, & \text{elsewhere} \end{cases} \tag{2.25}$$

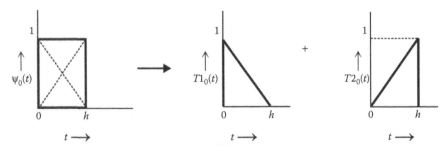

FIGURE 2.12
Dissection of the first member of a BPF set into two TFs.

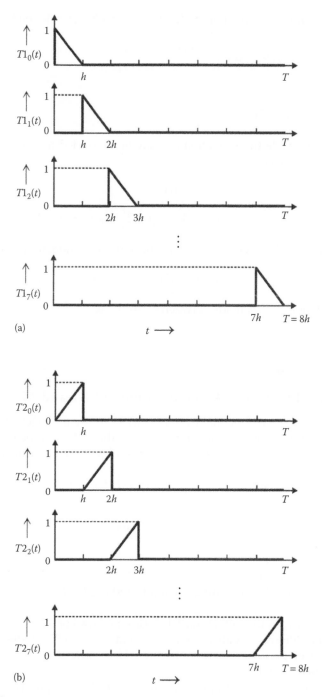

FIGURE 2.13
(a) A set of eight LHTF **T1**$_{(8)}(t)$; (b) a set of eight RHTF **T2**$_{(8)}(t)$.

and the $(i + 1)$th component of the RHTF vector $\mathbf{T2}_{(m)}(t)$ is defined as

$$T2_i(t) = \begin{cases} (t - ih)/h, & \text{for } ih \le t < (i+1)h \\ 0, & \text{elsewhere} \end{cases} \tag{2.26}$$

where:
$i = 0, 1, 2, \ldots, (m - 1)$

A square-integrable time function $f(t)$ may be expanded into an m-term TF series in $t \in [0, T]$ as

$$f(t) \approx \left[c_0\, c_1\, c_2 \ldots c_i \ldots c_{(m-1)} \right] \mathbf{T1}_{(m)}(t)$$
$$+ \left[d_0\, d_1\, d_2 \ldots d_i \ldots d_{(m-1)} \right] \mathbf{T2}_{(m)}(t) \underline{\Delta} \mathbf{C}^{\mathrm{T}} \mathbf{T1}_{(m)}(t) + \mathbf{D}^{\mathrm{T}} \mathbf{T2}_{(m)}(t) \tag{2.27}$$

The constant coefficients c_i's and d_i's in the above equation are given by

$$c_i \underline{\Delta} f(ih) \text{ and } d_i \underline{\Delta} f[(i+1)h] \tag{2.28}$$

and the relation $c_{i+1} = d_i$ holds between c_i's and d_i's.

2.10 Hybrid Function: A Combination of SHF and TF

Another new set of orthogonal functions was proposed by Deb et al. [49] in 2010. This new set was a linear combination of the SHFs and the RHTF set of the sets of TFs. The set obtained from such combination was termed "hybrid function" (HF) set [50–53].

To define an HF set, the $(i + 1)$th member $H_i(t)$ of the m-set HF $\mathbf{H}_{(m)}(t)$ is expressed as

$$H_i(t) = a_i S_i(t) + b_i T2_i(t) \tag{2.29}$$

where:
$i = 0, 1, 2, \ldots, (m - 1)$
a_i and b_i are weights or scaling constants and $0 \le t < T$

For convenience, in the following, $\mathbf{T}$ is written instead of $\mathbf{T2}$.

While the BPF set provides a staircase solution, the HF set always comes up with piecewise linear solution. Figure 2.14 illustrates how a function $f(t)$ is represented via HFs.

The function $f(t)$ is sampled at three equidistant points (sampling interval h) A, C, and E, with sample values c_0, c_1, and c_2. Now $f(t)$ can be

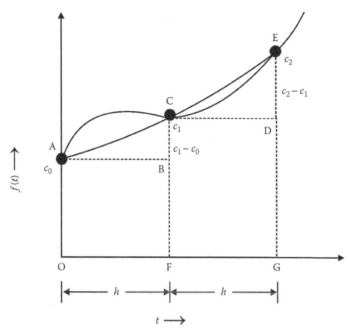

FIGURE 2.14
A function $f(t)$ represented via HFs.

expressed in a piecewise linear form by the two straight lines AC and CE, which are the sides of two adjacent trapeziums. Thus, expressing $f(t)$ via HFs, we can write

$$f(t) \approx H_0(t) + H_1(t)$$

$$= \{c_0 S_0(t) + (c_1 - c_0) T_0(t)\} + \{c_1 S_1(t) + (c_2 - c_1) T_1(t)\}$$

$$= \{c_0 S_0(t) + c_1 S_1(t)\} + \{(c_1 - c_0) T_0(t) + (c_2 - c_1) T_1(t)\}$$

$$= \mathbf{C}^{\mathrm{T}} \mathbf{S}_{(2)}(t) + \mathbf{D}^{\mathrm{T}} \mathbf{T}_{(2)}(t)$$

where:
$$\mathbf{C}^{\mathrm{T}} = [c_0 \ c_1]$$
$$\mathbf{D}^{\mathrm{T}} = [(c_1 - c_0) \ (c_2 - c_1)].$$

This set has been used for the applications such as identification [50], solution of differential equation [51–53]. Figure 2.15 shows a timescale history of the "alternative" class of orthogonal functions.

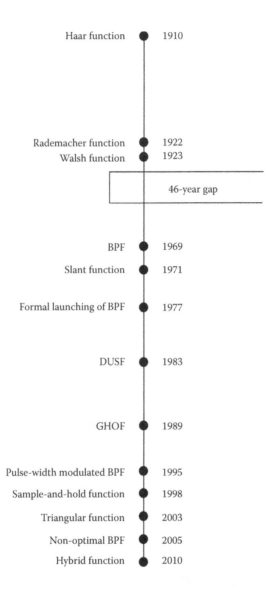

FIGURE 2.15
A timescale history of the "alternative" class of orthogonal functions. BPF, block pulse function; DUSF, delayed unit step function; GHOF, general hybrid orthogonal function.

2.11 Applications of Walsh Functions

By the 1930s, a firm foundation describing the mathematical properties of the Walsh functions was almost complete. But its application to the fields of engineering and technology had to wait till the discovery of semiconductors and digital computers. One could find the transparent connection between these two apparently wide-spaced branches of science: the binary connection. Scientists discovered the binary nature of Walsh functions, and its striking similarity to the familiar sine–cosine functions could entitle it for application efforts in some of the areas of science and technology. One immediate advantage was the task of analog multiplication. To multiply any signal by a Walsh function reduces the task to an appropriate sequence of sign changes that makes this usually difficult operation both simple and potentially accurate [54].

In the early 1960s, the first significant application of Walsh function in the field of communication was noted, and the credit goes mostly to Harmuth [55]. Consequently, Walsh function was found to be an efficient tool in the field of signal multiplexing. Several experimental multiplexing systems were developed, which made use of this nonsinusoidal technology.

The application of sequency techniques to communication problems has been particularly rewarding for those digital or two-level design processes in which the characteristics of the Walsh or other orthogonal sequency functions match closely the functional needs. In many cases, it has been found that certain characteristics of the orthogonal nonsinusoidal functions generate new ideas even in hardware development [56,57].

Harmuth [12], Taki and Hatori [58], and particularly Hübner [59] imparted the much-needed impetus through their outstanding achievements. They demonstrated some of the practical advantages of the newer techniques in the real working environment. These advantages include efficient multiplexing hardware, reduction of transmission bandwidth, and lowering of error rates.

Another important attribute of meaningful communication is the efficient coding of the transmitted data. Finally, the contribution of Harmuth [60] and others in opening up the new field of nonsinusoidal wideband radar communications is one of the major developments in sequency applications during the past decade.

The main stimulus to the use of sequency functions in signal processing applications lies in the ease with which their transformation can be implemented on a digital computer and, in some cases, the better matching of the function with the shape of the signal waveform being processed. The application of sequency digital processing technique covers a very wide field and includes various uses of spectral analysis. Some of these areas are seismic and other forms of low-frequency processing, radar and sonar applications, medical signal processing, speech processing, digital image processing, and so on.

The processing of fixed or changing visual images by using digital techniques requires the manipulation of multidimensional signal involving operations on large numbers of data values. Since this process generally involves high-speed on-line computer and substantial processing time, serious attempts are being made to find efficient alternative techniques. The sequency techniques—especially two-dimensional Walsh functions [12]—with their emphasis on rapid computational ease, play a significant role in these developments.

The traditional approach to the classification and design of digital logic systems is based on the table and diagram methods together with topological and algebraic operations [61]. These techniques, and indeed the general methods of Boolean algebra, are not now sufficient to deal with the more powerful logic systems made possible by using complementary metal–oxide–semiconductor (CMOS) and other large-scale integration (LSI) technology. A design philosophy that points a way out of this difficulty is the use of domain transformation for the proper logical description of digital designs. Intuitively, we recognize that bivalued functions such as those of Walsh and Rademacher are obvious candidates for the transformation process. There has been considerable development in applying these functions in many related fields [62–65] during the past three decades.

The widespread interest in practical applications of Walsh functions has stimulated further contributions to the mathematical theory. Of special interest is the logical differential calculus of Gibbs and Millard [66] and Gibbs [67]. In contrast to the sine–cosine functions, which often represent the characteristic solution to certain linear differential equations, the Walsh functions are shown to represent the solutions to what are known as the logical differential equations. Applications of the Gibbs derivative are found in mathematical logic [68], approximation theory [69], statistics [70,71], and linear system theory [72].

While these developments with the piecewise constant orthogonal functions were noted, a completely new vistas of analysis and application opened up in the field of systems and control. In 1973, Corrington [72] proposed a new technique for solving linear as well as nonlinear differential and integral equations with the help of Walsh functions. This gave an impetus to researchers, and a completely new band of application suitabilities of these functions was explored. In 1975, important technical papers relating Walsh functions to the field of systems and control were published. New ideas were proposed by Rao and Sivakumar [73–76], Rao and Palanisamy [77–79], Rao et al. [80], and Chen and Hsiao [18,81–84]. Other notable workers were Le Van et al. [85], Tzafestas [86], Chen and Shih [87,88], Chen and Lee [89], Chen [90], Mahapatra [91], Paraskevopoulos and Varoufakis [92], Moulden and Scott [93], and others.

The first positive step in Walsh domain analysis was the formulation of the operational matrix for integration. This was done independently by Rao and Sivakumar [73], Rao and Sivakumar [76], Chen and Hsiao [18,81–83], and Le Van et al. [85]. Le Van sensed that since the integral operator matrix had an inverse, the inverse must be the differential operator in the Walsh domain. However, he could not represent the general form of the operator matrix that was done by Chen and Hsiao [18] and Rao and Sivakumar [76].

Interestingly, the operational matrix for integration was first presented by Corrington [72] in the form of a table, but he failed to recognize the potentiality of the table as a matrix. This was first pointed out by C. F. Chen and his co-workers. They presented Walsh domain analysis with the operational matrices for integration as well as differentiation (1) to solve the problems of linear systems by state space model [18], (2) to design piecewise constant gains for optimal control [80], (3) for time domain synthesis [81], (4) to solve optimal control problem [82], (5) in variational problems [83], and (6) for fractional calculus as applied to distributed systems [20].

Rao used Walsh function for (1) system identification [73], (2) optimal control of time-delay systems [78], (3) identification of time-lag systems [74], (4) transfer function matrix identification in multiple-input, multiple-output (MIMO) systems [75], (5) parameter estimation [28], and (6) solving functional differential equations and related problems [28]. He first formulated operational matrices for stretch and delay [28,74,77–79]. He proposed a new technique for extension of computation beyond the limit of initial normal interval with the help of a "single-term Walsh series" approach [80] and estimated the error due to the use of different operational matrices [28].

Chen defined a "shift Walsh matrix" for solving delay differential equations [88] and used Walsh functions for parameter estimation of bilinear systems [87] as well as in the analysis of multidelay systems [90]. Paraskevopoulos determined the transfer function of a single-input, single-output (SISO) system from its impulse response with the help of Walsh functions and a fast Walsh algorithm [92]. Tzafestas [86] applied Walsh series approach for lumped and distributed system identification. Mahapatra [91] used Walsh functions for solving matrix Riccati equation arising in optimal control studies of linear diffusion equations. Moulden's work was concerned with the application of Walsh spectral analysis to ordinary differential equations in a very formal and a mathematical manner [93].

With the efforts of all these workers during the past decades, the mathematical basis of Walsh function methods has really become strong and versatile to encourage the application of the technique to analyze PE circuits and systems. From the study of different aspects of the Walsh functions, we find the following suitable qualities befitting with the analysis of PE systems:

1. Any member of the Walsh function set resembles to some extent the typical switching pattern of a PE converter, and the voltage output of such a converter could well be represented by Walsh functions in a simpler manner.
2. Walsh functions are defined in time domain. Thus, we do not need any inverse transformation.

3. The function set is complete and orthonormal, thereby offering the facility for easier mathematical manipulations, and it helps to design a fast computational algorithm with guaranteed minimum MISE.

4. The function is binary valued and is suitable for microprocessor or digital computer applications.

From the studies of Walsh functions and their applications, it seems worthwhile to carry out the analysis of PE systems in Walsh domain. To represent the load circuit suitably, we introduce the concept of a Walsh operational transfer function (WOTF) which, when multiplied with the input waveform, will yield the desired output variable. Definitely, the output variable would be in the form of a staircase waveform. But if a large number of basis Walsh functions are chosen to define a working set, the results could become more and more accurate.

It should be noted that BPFs could also serve our purpose and yield exactly the same results. But due to the ease of representation of some of the voltage waveforms of any PE converter system, we choose Walsh functions as the main vehicle along with the support from BPF as and when necessary for the analyses to follow.

References

1. Sansone, G., *Orthogonal Function*, Interscience Publishers Inc., New York, 1959.
2. Fourier, J. B., *Théorie analytique de la chaleur*, 1828. English Edition: *The Analytic Theory of Heat*, 1878. Reprinted by Dover Publication, New York, 1955.
3. Deb, A., Sen, S. K. and Datta, A. K., Walsh functions and their applications: A review, *IETE Tech. Rev.*, Vol. **9**, No. 3, pp. 238–252, 1992.
4. Kwong, C. P. and Chen, C. F., The convergence properties of block pulse series, *Int. J. Syst. Sci.*, Vol. **12**, pp. 745–751, 1981.
5. Haar, A., Zur theorie der orthogonalen funktionensysteme, *Math. Annal.*, Vol. **69**, pp. 331–371, 1910.
6. Ghosh, S., Deb, A. and Sarkar, G., On-Line block pulse implementation of a sine wave using microprocessor, *Measurement*, Vol. **45**, No. 6, pp. 1626–1632, 2012.
7. Shore, J. E., On application of Haar functions, *IEEE Trans. Commun.*, Vol. **21**, pp. 209–216, 1973.
8. Aboufadel, E. and Schlicker, S., *Discovering Wavelets*, Wiley Interscience, New York, 1999.
9. Burrus, C. S., Gopinath, R. A. and Guo, H., *Introduction to Wavelets and Wavelet Transforms*, Prentice-Hall, Upper Saddle River, NJ, 1998.
10. Rademacher, H., Einige sätze von allegemeinen orthogonal funktionen, *Math. Annal.*, Vol. **87**, pp. 122–138, 1922.
11. Walsh, J. L., A closed set of normal orthogonal functions, *Am. J. Math.*, Vol. **45**, pp. 5–24, 1923.

12. Harmuth, H. F., *Transmission of Information by Orthogonal Functions* (2nd Ed.), Springer-Verlag, Berlin, 1972.
13. Paley, R. E., A remarkable set of orthogonal functions, Proc. Lond. Math. Soc., Vol. **34**, pp. 241–279, 1932.
14. Henderson, K. W., Comment on 'computation of the fast Walsh–Fourier transform', *IEEE Trans. Comput.*, Vol. **C-19**, pp. 850–851, 1970.
15. Singmaster, D., *Notes on Rubik's Magic Cube*, Penguin Books, Great Britain, 1981.
16. Rao, K. R., Naramsimham, M. A. and Revulieri, K., Image data processing by Hadamard–Haar transform, *IEEE Trans. Comput.*, Vol. **C-24**, pp. 888–896, 1975.
17. Huang, D. M., Walsh–Hadamard–Haar hybrid transforms, IEEE Proceedings of the 5th International Conference on Pattern Recognition, Miami Beach, FL, December 1–14, 1980, pp. 180–182.
18. Chen, C. F. and Hsiao, C. H., A state space approach to Walsh series solution of linear systems, *Int. J. Syst. Sci.*, Vol. **6**, No. 9, pp. 833–858, 1975.
19. Wu, T. T., Chen, C. F. and Tsay, Y. T., Walsh operational matrices for fractional calculus and their applications to distributed systems, IEEE Symposium on Circuits and Systems, Munich, April 1976.
20. Chen, C. F., Tsay, Y. T. and Wu, T. T., Walsh operational matrices for fractional calculus and their applications to distributed systems, *J. Franklin Inst.*, Vol. **303**, No. 3, pp. 267–284, 1977.
21. Rao, G. P. and Srinivasan, T., Remarks on "author's reply" to "comments on design of piecewise constant gains for optimal control via Walsh functions," *IEEE Trans. Automat. Contr.*, Vol. **AC-23**, No. 4, pp. 762–763, 1978.
22. Deb, A., Sarkar, G. and Sen, S. K., Block pulse functions, the most fundamental of all piecewise constant basis functions, *Int. J. Syst. Sci.*, Vol. **25**, No. 2, pp. 351–363, 1994.
23. Sannuti, P., Analysis and synthesis of dynamic systems via block-pulse functions, *Proc. IEE*, Vol. **124**, No. 6, pp. 569–571, 1977.
24. Dalton, O. N., Further comments on "design of piecewise constant gains for optimal control via Walsh functions," *IEEE Trans. Automat. Contr.*, Vol. **AC-23**, No. 4, pp. 760–762, 1978.
25. Rao, G. P. and Srinivasan, T., Analysis and synthesis of dynamic systems containing time delays via block-pulse functions, *Proc. IEE*, Vol. **125**, No. 9, pp. 1064–1068, 1978.
26. Jan, Y. G. and Wong, K. M., Bilinear system identification by block pulse functions, *J. Franklin Inst.*, Vol. **312**, No. 5, pp. 349–359, 1981.
27. Chi-Hsu, W., On the generalization of block pulse operational matrices for fractional and operational calculus, *J. Franklin Inst.*, Vol. **315**, No. 2, pp. 91–102, 1983.
28. Rao, G. P., *Piecewise Constant Orthogonal Functions and Their Application to Systems and Control*, LNCIS-55, Springer-Verlag, Berlin, 1983.
29. Jiang, J. H. and Schaufelberger, W., *Block Pulse Functions and Their Application in Control System*, LNCIS-179, Springer-Verlag, Berlin, 1992.
30. Kalat, J. and Paraskevopoulos, P. N., Solution of multipoint boundary value problems via block-pulse functions, *J. Franklin Inst.*, Vol. **324**, No. 1, pp. 73–81, 1987.
31. Chen, W. L., Block pulse series analysis of scaled systems, *Int. J. Syst. Sci.*, Vol. **12**, No. 7, pp. 885–891, 1981.

32. Kung, F. C. and Chen, S. Y., Solution of integral equations using a set of block-pulse functions, *J. Franklin Inst.*, Vol. **306**, No. 4, pp. 283–291, 1978.

33. Rao, G. P. and Srinivasan, T., Multidimensional block-pulse functions and their use in the study of distributed parameter systems, *Int. J. Syst. Sci.*, Vol. **11**, No. 6, pp. 689–708, 1980.

34. Cheng, B. and Hsu, N. S., Analysis and parameter estimation of bilinear systems via block-pulse functions, *Int. J. Control*, Vol. **36**, No. 1, pp. 53–65, 1982.

35. Jiang, Z. H., New approximation method for inverse Laplace transforms using block-pulse functions, *Int. J. Syst. Sci.*, Vol. **18**, No. 10, pp. 1873–1888, 1987.

36. Enomoto, H. and Shibata, K., Orthogonal transform coding system for television signals, Proceedings of the Symposium on Application of Walsh Functions, Washington, DC, April, 1971, pp. 11–17.

37. Hwang, C., Solution of functional differential equation via delayed unit step functions, *Int. J. Syst. Sci.*, Vol. **14**, No. 9, pp. 1065–1073, 1983.

38. Patra, A. and Rao, G. P., General hybrid orthogonal functions and some potential applications in systems and control, *Proc. IEE Part D Control Theory Appl.*, Vol. **136**, No. 4, pp. 157–163, 1989.

39. Patra, A. and Rao, G. P., Continuous-time approach to self-tuning control: Algorithm, implementation and assessment, *Proc. IEE Part D Control Theory Appl.*, Vol. **136**, No. 6, pp. 333–340, 1989.

40. Patra, A. and Rao, G. P., *General Hybrid Orthogonal Functions and Their Applications in Systems and Control*, LNCIS-213, Springer, London, 1996.

41. Deb, A., Sarkar, G. and Sen, S. K., Linearly pulse-width modulated block pulse functions and their application to linear SISO feedback control system identification, *Proc. IEE Part D Control Theory Appl.*, Vol. **142**, No. 1, pp. 44–50, 1995.

42. Deb, A., Sarkar, G. and Sen, S. K., A new set of pulse-width modulated generalised block pulse functions (PWM-GBPF) and their application to cross/auto-correlation of time varying functions, *Int. J. Syst. Sci.*, Vol. **26**, No. 1, pp. 65–89, 1995.

43. Deb, A., Sarkar, G., Bhattacharjee, M. and Sen, S. K., A new set of piecewise constant orthogonal functions for the analysis of linear SISO systems with sample-and-hold, *J. Franklin Inst.*, Vol. **335B**, No. 2, pp. 333–358, 1998.

44. Deb, A., Sarkar, G. and Dasgupta, A., A complementary pair of orthogonal triangular function sets and its application to the analysis of SISO control systems, *J. Inst. Eng.*, Vol. **84**, pp. 120–129, 2003.

45. Deb, A., Sarkar, G. and Sengupta, A., *Triangular Orthogonal Functions for the Analysis of Continuous Time Systems*, Anthem Press, London, 2011.

46. Deb, A., Dasgupta, A. and Sarkar, G., A new set of orthogonal functions and its application to the analysis of dynamic systems, *J. Franklin Inst.*, Vol. **343**, pp. 1–26, 2006.

47. Ghosh, S., Deb, A., Choudhury, S. R. and Sarkar, G., Modeling and analysis of singular systems via orthogonal triangular functions, *J. Control Theory Appl.*, Vol. **11**, No. 2, pp. 141–148, 2013.

48. Deb, A., Ghosh, S., Choudhury, S. R. and Sarkar, G., A new recursive method for the analysis of linear time invariant dynamic systems via double-term triangular functions (DTTF) in state space environment, *J. Control Theory Appl.*, Vol. **11**, No. 1, pp. 108–115, 2013.

49. Deb, A., Ganguly, A., Biswas, A., Sarkar, G. and Mandal, P., Approximation and integration of time functions using a set of orthogonal hybrid functions (HF),

National Seminar in Electrical, Power Engineering, Electronics, and Computer (Tech Meet 2010), Pailan College of Management & Training, Kolkata, India, February 12–13, 2010, pp. 80–87.

50. Deb, A., Sarkar, G., Mandal, P., Biswas, A., Ganguly, A. and Biswas, D., Transfer function identification from impulse response via a new set of orthogonal hybrid functions (HF), *Appl. Math. Comput.*, Vol. **218**, No. 9, pp. 4760–4787, 2012.

51. Deb, A., Sarkar, G., Ganguly, A. and Biswas, A., Approximation, integration and differentiation of time functions using a set of orthogonal hybrid functions (HF) and their application to solution of first order differential equations, *Appl. Math. Comput.*, Vol. **218**, No. 9, pp. 4731–4759, 2012.

52. Deb, A., Sarkar, G., Ganguly, A. and Biswas, A., Numerical algorithm for the solution of third order differential equations in orthogonal hybrid function (HF) domain, India Conference INDICON 2011, Hyderabad, December 16–18, 2011.

53. Ganguly, A., Deb, A. and Sarkar, G., Numerical solution of second order linear differential equations using one-shot operational matrices in orthogonal hybrid function (HF) domain, Proceedings of the Conference ACODS 2012, Bangalore, February 16–18, 2012.

54. Hammond, J. L. and Johnson, R. S., A review of orthogonal square–wave functions and their application to linear networks, *J. Franklin Inst.*, Vol. **273**, pp. 211–225, 1962.

55. Harmuth, H. F., On the transmission of information by orthogonal time functions, *Trans. AIEE Commun. Electron.*, Vol. **79**, pp. 248–255, 1960.

56. Beauchamp, K. G., *Application of Walsh and Related Functions: With an Introduction to Sequency Theory*, Academic Press, London, 1984.

57. Beauchamp, K. G., *Walsh Functions and Their Applications*, Academic Press, London, 1975.

58. Taki, Y. and Hatori, M., P.C.M. communication system using Hadamard transformation, *Electron. Commun. Jpn.*, Vol. **49**, pp. 247–267, 1966.

59. Hübner, H., Analog and digital multiplexing by means of Walsh functions, Proceedings of the Symposium on Application on Walsh functions, Washington, DC, April, 1971, pp. 180–191.

60. Harmuth, H. F., *Nonsinusoidal Waves for Radar and Radio Communication*, Academic Press, New York, 1981.

61. Mano, M. M., *Computer Logic Design*, Prentice-Hall, Upper Saddle River, NJ, 1972.

62. Edwards, C. R., The application of the Rademacher–Walsh transform to Boolean function classification and threshold logic synthesis, *IEEE Trans. Comput.*, Vol. **C-24**, No. 1, pp. 48–62, 1975.

63. Edwards, C. R., A special class of universal logic gates and their evaluation under a Walsh transform, *Int. J. Electron.*, Vol. **44**, No. 1, pp. 49–59, 1978.

64. Bennetts, R. G. and Hurst, S. L., Rademacher–Walsh spectral transform: A new tool for problems in digital network fault diagnosis?, *Proc. IEE Part J. Comput. Digit. Technol.*, Vol. **1**, pp. 38–44, 1978.

65. Edwards, C. R., The application of the Rademacher-Walsh transform to digital circuit synthesis, Proceedings of the Theory and Application of Walsh Functions, Hatfield Polytechnic, England, 1973.

66. Gibbs, J. E. and Millard, M. J., Walsh functions as solutions of a logical differential equation, DES Report No. 1 & 2, National Physical Laboratory, England, 1969.

67. Gibbs, J. E., Sine waves and Walsh waves in physics, Proceedings of the Application of Walsh Functions, Washington, DC, March 31–April 3, 1970.
68. Liedl, P., Harmonische analysis bei aussagenkalkuelen, *Math. Logik*, Vol. **13**, pp. 158–167, 1970.
69. Pearl, J., Applications of Walsh transform to statistical analysis, *IEEE Trans. Syst. Man Cyber.*, Vol. **SMC-1**, pp. 111–119, 1971.
70. Nambiar, K. K., Approximation and representation of joint probability distribution of binary random variables by Walsh functions, Proceedings of the Symposium on Application of Walsh Functions, Washington, DC, March 31–April 3, 1970, pp. 70–72.
71. Pichler, F., Some aspects of a theory of correlation with respect to Walsh harmonic analysis, Report No. AD 714 596, Maryland University, College Park, MD, 1970.
72. Corrington, M. S., Solution of differential and integral equations with Walsh functions, *IEEE Trans. Circuit Theory*, Vol. **CT-20**, No. 5, pp. 470–476, 1973.
73. Rao, G. P. and Sivakumar, L., System identification via Walsh functions, *Proc. IEE*, Vol. **122**, No. 10, pp. 1160–1161, 1975.
74. Rao, G. P. and Sivakumar, L., Identification of time–lag systems via Walsh functions, *IEEE Trans. Automat. Contr.*, Vol. **AC-24**, No. 5, pp. 806–808, 1979.
75. Rao, G. P. and Sivakumar, L., Transfer function matrix identification in MIMO systems via Walsh functions, *Proc. IEEE*, Vol. **69**, No. 4, pp. 465–466, 1981.
76. Rao, G. P. and Sivakumar, L., Piecewise linear system identification via Walsh functions, *Int. J. Syst. Sci.*, Vol. **13**, No. 5, pp. 525–530, 1982.
77. Rao, G. P. and Palanisamy, K. R., A new operational–matrix for delay via Walsh functions and some aspects of its algebra and applications, Proceedings of the National Systems Conference, NSC–78, PAU Ludhiana, India, September 1978.
78. Rao, G. P. and Palanisamy, K. R., Optimal control of time-delay systems via Walsh functions, 9th IFIP Conference on Optimization Techniques, Polish Academy of Sciences, Systems Research Institute, Poland, September 1979.
79. Rao, G. P. and Palanisamy, K. R., Walsh stretch matrices and functional differential equations, *IEEE Trans. Automat. Contr.*, Vol. **27**, pp. 272–276, 1982.
80. Rao, G. P., Palanisamy, K. R. and Srinivasan, T., Extension of computation beyond the limit of initial normal interval in Walsh series analysis of dynamical systems, *IEEE Trans. Automat. Contr.*, Vol. **AC-25**, No. 2, pp. 317–319, 1980.
81. Chen, C. F. and Hsiao, C. H., Design of piecewise constant gains for optimal control via Walsh functions, *IEEE Trans. Automat. Contr.*, Vol. **AC-20**, No. 5, pp. 596–603, 1975.
82. Chen, C. F. and Hsiao, C. H., Time domain synthesis via Walsh functions, *Proc. IEE*, Vol. **122**, No. 5, pp. 565–570, 1975.
83. Chen, C. F. and Hsiao, C. H., Walsh series analysis in optimal control, *Int. J. Control*, Vol. **21**, No. 6, pp. 881–897, 1975.
84. Chen, C. F. and Hsiao, C. H., A Walsh series direct method for solving variational problems, *J. Franklin Inst.*, Vol. **300**, No. 4, pp. 265–280, 1975.
85. Le Van, T., Tam, L. D. C. and Van Houtte, N., On direct algebraic solutions of linear differential equations using Walsh transforms, *IEEE Trans. Circuits Syst.*, Vol. **CAS-22**, No. 5, pp. 419–422, 1975.
86. Tzafestas, S. G. (Ed.), *Walsh Functions in Signal and Systems Analysis and Design*, Van Nostrand Reinhold, New York, 1985.

87. Chen, W. L. and Shih, Y. P., Parameter estimation of bilinear systems via Walsh functions, *J. Franklin Inst.*, Vol. **305**, No. 5, pp. 249–257, 1978.
88. Chen, W. L. and Shih, Y. P., Shift Walsh matrix and delay differential equations, *IEEE Trans. Automat. Contr.*, Vol. **AC-23**, No. 6, pp. 1023–1028, 1978.
89. Chen, W. L. and Lee, C. L., Walsh series expansions of composite functions and its applications to linear systems, *Int. J. Syst. Sci.*, Vol. **13**, No. 2, pp. 219–226, 1982.
90. Chen, W. L., Walsh series analysis of multi-delay systems, *J. Franklin Inst.*, Vol. **313**, No. 4, pp. 207–217, 1982.
91. Mahapatra, G. B., Solution of optimal control problem of linear diffusion equation via Walsh functions, *IEEE Trans. Automat. Contr.*, Vol. **AC-25**, No. 2, pp. 319–321, 1980.
92. Paraskevopoulos, P. N. and Varoufakis, S. J., Transfer function determination from impulse response via Walsh functions, *Int. J. Circuit Theory Appl.*, Vol. **8**, pp. 85–89, 1980.
93. Moulden, T. H. and Scott, M. A., Walsh spectral analysis for ordinary differential equations: Part 1—initial value problems, *IEEE Trans. Circuits Syst.*, Vol. **CAS-35**, No. 6, pp. 742–745, 1988.

3

Walsh Domain Operational Method of System Analysis

In this chapter, the utilities of Walsh functions in analyzing linear time-invariant single-input, single-output (SISO) systems using a new operational technique are presented. This technique has some definite advantages, discussed below. In addition, some interesting characteristics that indicate computational instability of the Walsh function technique have also been investigated. Despite such a disadvantage, the presented method has significant advantages over the traditional Laplace domain technique in the form that

1. It avoids any kind of inverse transformation always associated with the Laplace domain analysis.
2. It is more compatible to digital framework and computer manipulation.
3. It can represent input square wave functions quite elegantly, and such functions can be represented by Walsh functions effortlessly almost by inspection in many cases.

The method utilizes a Walsh operational transfer function (WOTF) [1] for analyzing SISO systems, and time-scaled operational matrices are defined to obtain time-scaled WOTFs.

3.1 Introduction to Operational Matrices

Corrington [2] first constructed the Walsh table using Walsh functions for solving differential as well as integral equations. He observed that in each small subinterval of time, the Walsh domain approximate solution approaches the mean value of the true solution. The solution technique is simple due to the fact that Walsh functions have simple rules for multiplication. Also, the computation of partial sums of the Walsh series to obtain a staircase solution is relatively easy.

Corrington formed a table of coefficients for integrating Walsh functions, but he did not realize the potentiality of the table of coefficients as a matrix. This was first indicated by Chen and Hsiao [3] who presented the table in the form of an operational matrix for integer integration and used the same for solving linear systems by state-space approach.

However, more or less concurrently, such operational technique was independently presented by Rao and Sivakumar [4] and Le Van et al. [5] for solving linear differential equations and system identification. Later, Chen and his coworkers used similar techniques for time-domain synthesis [6], solving optimal control problems [7], and solution of distributed systems [8].

3.1.1 Operational Matrix for Integration

It has been shown earlier that any function $f(t)$, which is square integrable in the interval [0, 1), can be expanded into a Walsh series in a manner equivalent to the well-known Fourier series expansion. This may be expressed according to Equation 2.1 as follows [6,7]:

$$f(t) = \sum_{n=0}^{\infty} c_n \phi_n(t) = c_0 \phi_0 + c_1 \phi_1 + c_2 \phi_2 + \cdots + c_n \phi_n + \cdots \tag{3.1}$$

where:

$c_0, c_1, c_2, \ldots, c_n$ are the coefficients of the Walsh series to be determined by

$$c_n = \int_0^1 \phi_n(t) f(t) dt \tag{3.2}$$

and

$\phi_0, \phi_1, \phi_2, \ldots, \phi_n$ are Walsh functions in the dyadic order

Equation 3.2 is chosen in such a way that the integral square error given by

$$\lim_{N \to \infty} \int_0^1 \left[f(t) - \sum_0^N c_n \phi_n(t) \right]^2 dt = 0 \tag{3.3}$$

is minimized. Obviously, the more the number of coefficients, the more accurate is the approximation.

As seen from Figure 2.5, the first Walsh function of the Walsh series, $\phi_0(t)$, is a unit step function. When this function is integrated, it yields a ramp function $f(t) = t$. Integration of other Walsh functions, $\phi_1(t)$, $\phi_2(t)$, and so on, generates various triangular waves. These triangular functions can also be expanded in a similar fashion into Walsh series, giving staircase approximation for each of them. Figure 3.1 shows the first four Walsh functions and their first integrals.

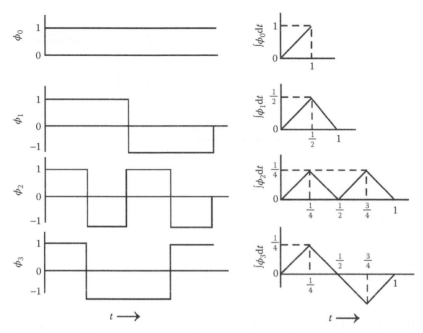

FIGURE 3.1
First four Walsh functions and their first integrals. (Reprinted from Chen, C. F. and Hsiao, C. H., *Int. J. Syst. Sci.*, 6, 833–858, 1975. With permission.)

Thus,

$$\int \phi_0(t)dt = f(t) = t$$

Let us expand this function in $t \in [0,1)$ using four Walsh functions.
Equation 3.2 yields the Walsh coefficients as

$$\left.\begin{aligned}
c_0 &= \int_0^1 \phi_0(t)\, t\, dt = \frac{1}{2} \\
c_1 &= \int_0^1 \phi_1(t)\, t\, dt = -\frac{1}{4} \\
c_2 &= \int_0^1 \phi_2(t)\, t\, dt = -\frac{1}{8} \\
c_3 &= \int_0^1 \phi_3(t)\, t\, dt = 0
\end{aligned}\right\} \tag{3.4}$$

Therefore, $t \approx (1/2)\phi_0(t) - (1/4)\phi_1(t) - (1/8)\phi_2(t) + (0)\phi_3(t)$ (Program B.3 in Appendix B).

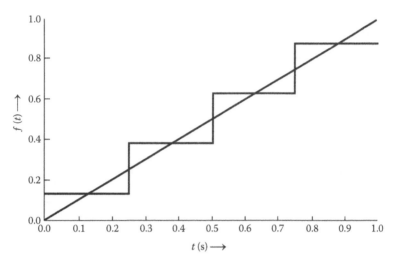

FIGURE 3.2
Walsh domain approximation of a unit ramp function for $m = 4$ and $T = 1$ second.

Thus, the ramp function is approximated in the form of a staircase wave shown in Figure 3.2. Equation 3.4 may be written in the following matrix form:

$$\int \phi_0(t)dt \approx \begin{bmatrix} \dfrac{1}{2} & -\dfrac{1}{4} & -\dfrac{1}{8} & 0 \end{bmatrix} \begin{bmatrix} \phi_0(t) \\ \phi_1(t) \\ \phi_2(t) \\ \phi_3(t) \end{bmatrix} = c_0 \Phi_{(4)} \qquad (3.5)$$

Similarly, other Walsh functions $\phi_1(t)$, $\phi_2(t)$, $\phi_3(t)$, and $\phi_4(t)$ are integrated and then expanded via Walsh series.

Therefore,

$$\int \phi_1(t)dt \approx \begin{bmatrix} \dfrac{1}{4} & 0 & 0 & -\dfrac{1}{8} \end{bmatrix} \begin{bmatrix} \phi_0(t) \\ \phi_1(t) \\ \phi_2(t) \\ \phi_3(t) \end{bmatrix} = c_1 \Phi_{(4)}$$

$$\int \phi_2(t)dt \approx \begin{bmatrix} \dfrac{1}{8} & 0 & 0 & 0 \end{bmatrix} \begin{bmatrix} \phi_0(t) \\ \phi_1(t) \\ \phi_2(t) \\ \phi_3(t) \end{bmatrix} = c_2 \Phi_{(4)}$$

$$\int \phi_3(t)dt \approx \begin{bmatrix} 0 & \dfrac{1}{8} & 0 & 0 \end{bmatrix} \begin{bmatrix} \phi_0(t) \\ \phi_1(t) \\ \phi_2(t) \\ \phi_3(t) \end{bmatrix} = c_3 \Phi_{(4)}$$

Thus, we can write the relationship between Walsh functions and their first integrals in the matrix form as

$$\begin{bmatrix} \int \phi_0(t)dt \\ \int \phi_1(t)dt \\ \int \phi_2(t)dt \\ \int \phi_3(t)dt \end{bmatrix} \approx \begin{bmatrix} \dfrac{1}{2} & -\dfrac{1}{4} & -\dfrac{1}{8} & 0 \\ \dfrac{1}{4} & 0 & 0 & -\dfrac{1}{8} \\ \dfrac{1}{8} & 0 & 0 & 0 \\ 0 & \dfrac{1}{8} & 0 & 0 \end{bmatrix} \begin{bmatrix} \phi_0(t) \\ \phi_1(t) \\ \phi_2(t) \\ \phi_3(t) \end{bmatrix} \tag{3.6}$$

or in compact form,

$$\int \Phi_{(4)}(t)dt \approx G_{(4)}\Phi_{(4)}(t) \tag{3.7}$$

where the subscript denotes the dimension of the Walsh series as well as that of the square operational matrix G.

The square matrix G is known as the operational matrix for integration because it performs integration on $\Phi(t)$ when it is multiplied by $\Phi(t)$ in Equation 3.7. It is chosen as a square matrix of order $m = 2^p, p = 1, 2, 3,\ldots$ (the number of component functions chosen).

Equation 3.7 is obviously an approximate formula, and its accuracy depends on the dimension of Φ or G.

Now, we can determine the operational matrix for integration of larger dimensions in a similar fashion. Figure 3.3 shows the first eight Walsh functions and their first integrals.

Integration of $\phi_0(t)$ can be expanded into a Walsh series of eight terms with the help of Equations 3.1 and 3.2 (Program B.4 in Appendix B):

$$\int \phi_0(t)dt = f(t) = t \approx \frac{1}{2}\phi_0(t) - \frac{1}{4}\phi_1(t) - \frac{1}{8}\phi_2(t) - \frac{1}{16}\phi_4(t) \tag{3.8}$$

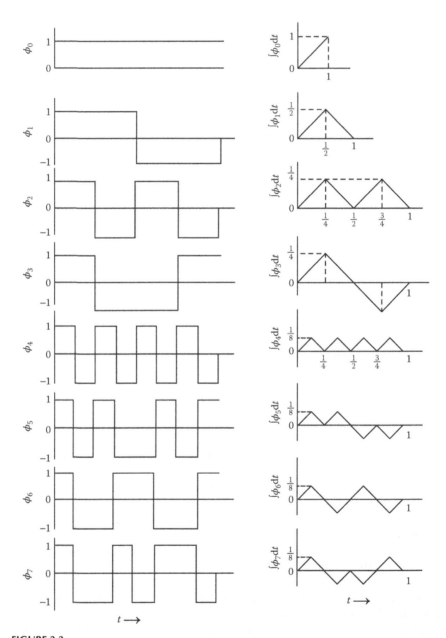

FIGURE 3.3
First eight Walsh functions and their first integrals. (Reprinted from Chen, C. F. and Hsiao, C. H., *Int. J. Syst. Sci.*, 6, 833–858, 1975. With permission.)

Equation 3.8 may be written in the following matrix form:

$$\int \phi_0(t) dt \approx \begin{bmatrix} \dfrac{1}{2} & -\dfrac{1}{4} & -\dfrac{1}{8} & 0 & -\dfrac{1}{16} & 0 & 0 & 0 \end{bmatrix} \begin{bmatrix} \phi_0(t) \\ \phi_1(t) \\ \phi_2(t) \\ \phi_3(t) \\ \phi_4(t) \\ \phi_5(t) \\ \phi_6(t) \\ \phi_7(t) \end{bmatrix} = \mathbf{c_0}\,\Phi_{(8)} \qquad (3.9)$$

Figure 3.4 (Program B.4 in Appendix B) shows Walsh approximation of the unit ramp function using Equation 3.9.

Integration of other Walsh functions, $\phi_1(t)$, $\phi_2(t)$, ..., $\phi_7(t)$, which generates various triangular waves, can also be expanded in a similar fashion into Walsh series, giving staircase approximation for each triangular function.

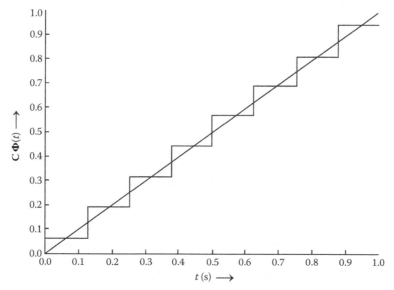

FIGURE 3.4
Walsh domain approximation of the unit ramp function for $m = 8$ and $T = 1$ second.

Considering the first eight functions of the Walsh series, the Walsh functions and their corresponding first integrations could be related by the matrix equation having a form similar to Equation 3.6. This is expressed in the following equation:

$$
\begin{bmatrix}
\int \phi_0(t)dt \\
\int \phi_1(t)dt \\
\int \phi_2(t)dt \\
\int \phi_3(t)dt \\
\int \phi_4(t)dt \\
\int \phi_5(t)dt \\
\int \phi_6(t)dt \\
\int \phi_7(t)dt
\end{bmatrix}
\approx
\begin{bmatrix}
\frac{1}{2} & -\frac{1}{4} & -\frac{1}{8} & 0 & -\frac{1}{16} & 0 & 0 & 0 \\
\frac{1}{4} & 0 & 0 & -\frac{1}{8} & 0 & -\frac{1}{16} & 0 & 0 \\
\frac{1}{8} & 0 & 0 & 0 & 0 & 0 & -\frac{1}{16} & 0 \\
0 & \frac{1}{8} & 0 & 0 & 0 & 0 & 0 & -\frac{1}{16} \\
\frac{1}{16} & 0 & 0 & 0 & 0 & 0 & 0 & 0 \\
0 & \frac{1}{16} & 0 & 0 & 0 & 0 & 0 & 0 \\
0 & 0 & \frac{1}{16} & 0 & 0 & 0 & 0 & 0 \\
0 & 0 & 0 & \frac{1}{16} & 0 & 0 & 0 & 0
\end{bmatrix}
\begin{bmatrix}
\phi_0(t) \\
\phi_1(t) \\
\phi_2(t) \\
\phi_3(t) \\
\phi_4(t) \\
\phi_5(t) \\
\phi_6(t) \\
\phi_7(t)
\end{bmatrix}
\tag{3.10}
$$

Equation 3.10 may be written in the following compact form of a vector–matrix equation:

$$
\int \Phi_{(8)}(t)dt \approx G_{(8)}\Phi_{(8)}(t)
\tag{3.11}
$$

The dimension of $\Phi(t)$ or G indicates that eight-component Walsh functions have been used and the interval 0 to 1 has been divided into eight equal subintervals. Thus, G is a square matrix of order eight, and hence mathematical manipulations become easy.

It is interesting to note that the upper-left quarter of $G_{(8)}$ is exactly $G_{(4)}$. The upper-right quarter is a diagonal matrix and the lower-left corner matrix is also a diagonal matrix, while the lower-right corner (4×4) matrix is a null matrix.

Hence, if the operational matrix for integration $G_{(16)}$ is derived, we will see that the upper-left quarter matrix is exactly $G_{(8)}$. Similar partitioning of $G_{(8)}$ will produce $G_{(4)}$.

The general operational matrix for integration, as given by Chen and Hsiao [3], is

$$
\mathbf{G}_{(m)} = \begin{bmatrix}
\begin{array}{c|cc|c}
\begin{matrix} 1/2 \\ \end{matrix} & \multicolumn{2}{c|}{-(2/m)\,\mathbf{I}_{(m/8)}\;\; -(1/m)\,\mathbf{I}_{(m/4)}} & \\ \hline
(2/m)\,\mathbf{I}_{(m/8)} & \mathbf{0}_{(m/8)} & & -(1/2m)\,\mathbf{I}_{(m/2)} \\ \hline
(1/m)\,\mathbf{I}_{(m/4)} & & \mathbf{0}_{(m/4)} & \\
\multicolumn{2}{c|}{(1/2m)\,\mathbf{I}_{(m/2)}} & & \mathbf{0}_{(m/2)}
\end{array}
\end{bmatrix}
\tag{3.12}
$$

which is of order $(m \times m)$.

This pattern is basically of repetitive nature. Its characteristics enable us to construct the operational matrices of larger dimensions.

3.1.1.1 Representation of Integration of a Function Using Operational Matrix for Integration

Example 3.1

Consider the function $f_1(t) = \sin(\pi t)$ (Program B.5 in Appendix B).

For $m = 4$ and $T = 1$ second, the function may be represented in the Walsh domain as

$$
f_1(t) \approx [0.6366 \quad 0 \quad 0 \quad -0.2637]\Phi_{(4)}(t) \triangleq \mathbf{C}\,\Phi_{(4)}(t)
$$

Using the operational matrix $\mathbf{G}_{(4)}$ from Equation 3.6, we can compute the integration of the function $f_1(t)$ in the Walsh domain.

Let $f_2(t) = \int_0^t f_1(t)\,dt$
Then,

$$
f_2(t) \approx \bar{f}_2(t) = \mathbf{C\,G}\,\Phi_{(4)}(t)\,dt = [0.6366 \quad 0 \quad 0 \quad -0.2637]\,\mathbf{G}_{(4)}\;\Phi_{(4)}(t)
$$

or
$$\tag{3.13}$$

$$
\bar{f}_2(t) = [0.3183 \quad -0.1921 \quad -0.0796 \quad 0]\Phi_{(4)}(t)
$$

Using exact integration, we get $f_2(t) = (1/\pi)(1 - \cos \pi t)$.

Now we expand $f_2(t)$ directly in the Walsh domain to assess the efficiency of the operational matrix $\mathbf{G}$. Using this direct expansion $\hat{f}_2(t)$, we have

$$
\hat{f}_2(t) = [0.3183 \quad -0.2026 \quad -0.0839 \quad 0]\Phi_{(4)}(t)
\tag{3.14}
$$

Table 3.1 shows the comparison between the Walsh coefficients solved using the operational matrix and the exact Walsh coefficients to give an idea about the efficiency of operational matrix $\mathbf{G}$.

TABLE 3.1

Comparison between the Walsh Coefficients Solved Using Operational Matrix and the Walsh Coefficients Obtained through Direct Expansion for $m = 4$ and $T = 1$ second

Walsh Coefficients Determined Using Operational Matrix (c_i')	Walsh Coefficients Obtained through Direct Expansion (c_i)	$\text{Error} = \dfrac{c_i - c_i'}{c_i} \times 100\ (\%)$
0.3183	0.3183	0
−0.1921	−0.2026	5.1941
−0.0796	−0.0839	5.1941
0	0	−

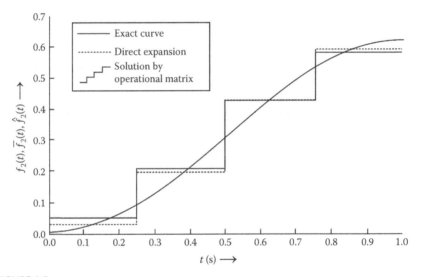

FIGURE 3.5

Integration of the function $f_1(t)$ using the operational matrix compared with the direct expansion of the result $f_2(t)$ in the Walsh domain for $m = 4$ and $T = 1$ second along with the exact curve.

In addition, from Equation 2.4, it is found that the mean integral square error (MISE) for the Walsh approximation using operational matrix for integration is 0.0027, whereas the MISE for the Walsh approximation via direct expansion is 0.0026. Thus, the error in the operational matrix-based integration technique proves to be very small.

Figure 3.5 compares the Walsh approximations (using operational matrix and direct expansion) with the exact curve (Program B.5 in Appendix B).

For $m = 8$ and $T = 1$ second, the function may be represented more accurately in the Walsh domain as

$$f_1(t) \approx [0.6366\ 0\ 0\ -0.2637\ 0\ -0.1266\ -0.0525\ 0]\Phi_{(8)}(t) \triangleq \mathbf{C}\,\Phi_{(8)}(t) \quad (3.15)$$

Using the operational matrix $\mathbf{G}_{(8)}$ from Equation 3.10, we can compute the integration of the function $f_1(t)$ in the Walsh domain with improved accuracy.

Let $f_2(t) = \int_0^t f_1(t)\,dt$

Using Equations 3.10 and 3.15, we get

$$f_2(t) \approx \overline{f}_2(t) = \mathbf{C}\,\mathbf{G}\,\Phi_{(8)}(t)\,dt$$

$$= [0.6366\ 0\ 0\ -0.2637\ 0\ -0.1266\ -0.0525\ 0]\mathbf{G}_{(8)}\,\Phi_{(8)}(t)$$

(3.16)

or

$$\overline{f}_2(t) = [0.3183\ -0.2000\ -0.0829\ 0\ -0.0398\ 0\ 0\ 0.0165]\Phi_{(8)}(t)$$

Now we expand $f_2(t)$ directly in the Walsh domain for $m = 8$ and $T = 1$ second to assess the efficiency of the operational matrix $\mathbf{G}$. Using this direct expansion $\hat{f}_2(t)$, we have

$$\hat{f}_2(t) = [0.3183\ -0.2026\ -0.0839\ 0\ -0.0403\ 0\ 0\ 0.0167]\Phi_{(8)}(t) \quad (3.17)$$

Figure 3.6 (Program B.6 in Appendix B) represents Walsh approximations with the exact curve. Table 3.2 compares the results obtained by two different methods.

In addition, it is found that the MISE for Walsh approximation using the operational matrix is 0.000656, whereas the MISE for direct Walsh expansion is 0.000647. The error is very negligible, and the MISE in the former case will be reduced further if m is increased.

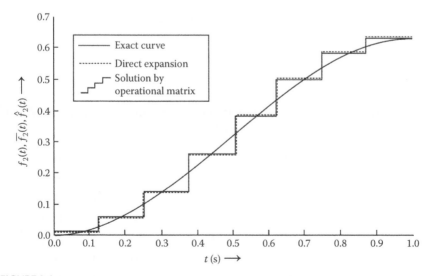

FIGURE 3.6

Integration of the function $f_1(t)$ using the operational matrix compared with the direct expansion of the result $f_2(t)$ in the Walsh domain for $m = 8$ and $T = 1$ second along with the exact curve.

TABLE 3.2

Comparison between the Walsh Coefficients Solved Using Operational Matrix and the Walsh Coefficients Obtained through Direct Expansion for $m = 8$ and $T = 1$ second

Walsh Coefficients Determined Using Operational Matrix (c_i')	Walsh Coefficients Obtained through Direct Expansion (c_i)	Error $= \dfrac{c_i - c_i'}{c_i} \times 100\ (\%)$
0.3183	0.3183	0
−0.2000	−0.2026	1.2884
−0.0829	−0.0839	1.2884
0	0	–
−0.0398	−0.0403	1.2884
0	0	–
0	0	–
0.0165	0.0167	1.2884

3.1.2 Operational Matrix for Differentiation

The operational matrix for differentiation cannot be obtained by differentiating Walsh functions as was done by integration in the case of integration matrix because differentiation of Walsh functions gives rise to delta functions and their higher derivatives, which could not be expanded into a Walsh series having a finite number of terms. This problem is solved indirectly by inverting the operational matrix $\mathbf{G}_{(4)}$, which is $\mathbf{D}_{(4)}$—the operational matrix for differentiation [8] in the Walsh domain. Hence,

$$\mathbf{D}_{(4)} = \left[\mathbf{G}_{(4)}\right]^{-1} = \begin{bmatrix} 0 & 0 & 8 & 0 \\ 0 & 0 & 0 & 8 \\ -8 & 0 & 32 & -16 \\ 0 & -8 & 16 & 0 \end{bmatrix} \qquad (3.18)$$

and also, we can find the operational matrix for differentiation in the Walsh domain having larger dimensions such as

$$\mathbf{D}_{(8)} = \left[\mathbf{G}_{(8)}\right]^{-1} = \begin{bmatrix} 0 & 0 & 0 & 0 & 16 & 0 & 0 & 0 \\ 0 & 0 & 0 & 0 & 0 & 16 & 0 & 0 \\ 0 & 0 & 0 & 0 & 0 & 0 & 16 & 0 \\ 0 & 0 & 0 & 0 & 0 & 0 & 0 & 16 \\ -16 & 0 & 0 & 0 & 128 & -64 & -32 & 0 \\ 0 & -16 & 0 & 0 & 64 & 0 & 0 & -32 \\ 0 & 0 & -16 & 0 & 32 & 0 & 0 & 0 \\ 0 & 0 & 0 & -16 & 0 & 32 & 0 & 0 \end{bmatrix} \qquad (3.19)$$

The general formula for the operational matrix for differentiation, as given by Chen and Hsiao [7], is as follows:

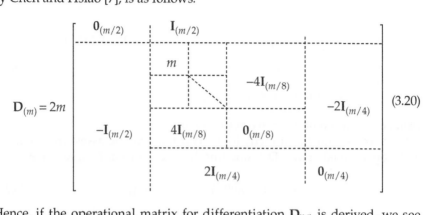

$$D_{(m)} = 2m \begin{bmatrix} 0_{(m/2)} & I_{(m/2)} & & \\ & m & -4I_{(m/8)} & -2I_{(m/4)} \\ -I_{(m/2)} & 4I_{(m/8)} & 0_{(m/8)} & \\ & 2I_{(m/4)} & & 0_{(m/4)} \end{bmatrix} \qquad (3.20)$$

Hence, if the operational matrix for differentiation $D_{(16)}$ is derived, we see that the upper-left corner is exactly $D_{(8)}$. Subsequently, by partitioning $D_{(8)}$, $D_{(4)}$ can be obtained. Thus, it has a similar repetitive nature like operational matrix for integration G.

It is observed that there is marked similarity between the Walsh differential operator $D_{(m)}$ and the Laplace operator s. That means, the operational matrix $D_{(m)}$ performs like a differentiator in the Walsh domain, just like the operator s in the Laplace domain.

Similarly, the operational matrix for integration $G_{(m)}$, given by Equation 3.12, performs like an integrator in the Walsh domain, as the operator s^{-1} does in the Laplace domain.

These important observations help to build the concept of **WOTF**, which is employed for the analysis of linear systems in the Walsh domain. However, before elaborating on the philosophy of the new Walsh domain operational technique of system analysis, we present the time-scaling properties of the operational matrices $G_{(m)}$ and $D_{(m)}$. For obtaining time-domain solutions of systems within a definite time interval, the time-scaling transformation is very necessary.

3.1.2.1 Representation of Differentiation of a Function Using Operational Matrix for Differentiation

Example 3.2

Consider the function $f_1(t) = \sin(\pi t)$.

For $m = 4$ and $T = 1$ second, the function may be represented in the Walsh domain as

$$f_1(t) \approx [0.6366 \quad 0 \quad 0 \quad -0.2637]\Phi_{(4)}(t) \triangleq C\,\Phi_{(4)}(t) \qquad (3.21)$$

Using the operational matrix $D_{(4)}$ from Equation 3.18, we can compute the differentiation of the function $f_1(t)$ in the Walsh domain.

Let $f_2(t) = \left(\dfrac{d}{dt}\right) f_1(t)$

Then,

$$f_2(t) \approx \bar{f}_2(t) = \mathbf{C} \, \mathbf{D} \, \Phi_{(4)}(t) \, dt = [0.6366 \quad 0 \quad 0 \quad -0.2637] \, \mathbf{D}_{(4)} \, \Phi_{(4)}(t)$$

or   (3.22)

$$\bar{f}_2(t) = [0 \quad 2.1096 \quad 0.8738 \quad 0] \Phi_{(4)}(t)$$

Using exact differentiation, we get $f_2(t) = \pi \cos \pi t$.

Now, we expand $f_2(t)$ directly in the Walsh domain to assess the accuracy of the operational matrix **D**. Using this direct expansion $\hat{f}_2(t)$, we have

$$\hat{f}_2(t) = [0 \quad 2.0000 \quad 0.8284 \quad 0] \Phi_{(4)}(t)$$ (3.23)

Figure 3.7 shows the comparison between Walsh approximations with the exact solution (Program B.7 in Appendix B).

Table 3.3 shows the comparison between Walsh coefficients solved using operational matrix and Walsh coefficients obtained via direct expansion to get an idea about the accuracy of the operational matrix **D**.

In addition, using Equation 2.4, it is noted that the MISE for Walsh approximation using **D** matrix is 0.2626, whereas the MISE for direct Walsh expansion of the differentiated function is 0.2485. The MISE 0.2626 will evidently become smaller with increased m.

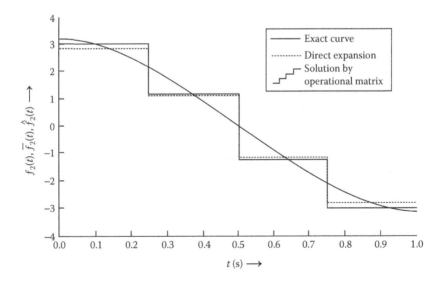

FIGURE 3.7
Differentiation of the function $f_1(t)$ using the operational matrix compared with the direct expansion of the result $f_2(t)$ in the Walsh domain for $m = 4$ and $T = 1$ second along with the exact curve.

TABLE 3.3

Comparison between the Walsh Coefficients Solved Using Operational Matrix and the Walsh Coefficients Obtained through Direct Expansion for $m = 4$ and $T = 1$ second

Walsh Coefficients Obtained Using Operational Matrix (c_i')	Walsh Coefficients Obtained through Direct Expansion (c_i)	$\text{Error} = \dfrac{c_i - c_i'}{c_i} \times 100$ (%)
0	0	–
2.1096	2.0000	−5.4786
0.8738	0.8284	−5.4786
0	0	–

For $m = 8$ and $T = 1$ second, the function may be represented in the Walsh domain as

$$f_1(t) \approx [0.6366 \quad 0 \quad 0 \quad -0.2637 \quad 0 \quad -0.1266 \quad -0.0525 \quad 0]\Phi_{(8)}(t)$$

$$\triangleq \mathbf{C}\,\Phi_{(8)}(t) \tag{3.24}$$

Using the operational matrix $\mathbf{D}_{(8)}$ from Equation 3.19, we can compute the differentiation of the function $f_1(t)$ in the Walsh domain with improved accuracy.

Let $f_2(t) = (d/dt)f_1(t)$

Then,

$$f_2(t) \approx \overline{f}_2(t) = \mathbf{C}\,\mathbf{D}\,\Phi_{(8)}(t)\,dt$$

$$= [0.6366 \quad 0 \quad 0 \quad -0.2637 \quad 0 \quad -0.1266 \quad -0.0525 \quad 0]\mathbf{D}_{(8)}\,\Phi_{(8)}(t) \tag{3.25}$$

or

$$\overline{f}_2(t) = [0 \quad 2.0261 \quad 0.8392 \quad 0 \quad 0.4030 \quad 0 \quad 0 \quad -0.1669]\Phi_{(8)}(t)$$

Using exact differentiation, we get $f_2(t) = \pi \cos \pi t$.

Now, we expand $f_2(t)$ directly in the Walsh domain to compute the Walsh coefficients c_i's and assess the accuracy of the operational matrix $\mathbf{D}$. Using this direct expansion $\hat{f}_2(t)$, we have

$$\hat{f}_2(t) = [0 \quad 2.0000 \quad 0.8284 \quad 0 \quad 0.3978 \quad 0 \quad 0 \quad -0.1648]\Phi_{(8)}(t) \tag{3.26}$$

Similar to Table 3.3, Table 3.4 presents a comparative study between the Walsh coefficients obtained by the two methods for $m = 8$.

Figure 3.8 (Program B.8 in Appendix B) compares the Walsh approximations along with the exact curve.

In addition, from Equation 2.4, it is observed that the MISE for Walsh approximation solved using matrix $\mathbf{D}$ is 0.0639, whereas the MISE for Walsh approximation via direct expansion is 0.0631. The MISE in the former case will be reduced still further if m is increased.

TABLE 3.4

Comparison between the Walsh Coefficients Determined Using Operational Matrix and the Walsh Coefficients Obtained through Direct Expansion for $m = 8$ and $T = 1$ second

Walsh Coefficients Determined Using Operational Matrix (c_i')	Walsh Coefficients Obtained through Direct Expansion (c_i)	$\text{Error} = \dfrac{c_i - c_i'}{c_i} \times 100\ (\%)$
0	0	–
2.0261	2.0000	−1.3052
0.8392	0.8284	−1.3052
0	0	–
0.4030	0.3978	−1.3052
0	0	–
0	0	–
−0.1669	−0.1648	−1.305

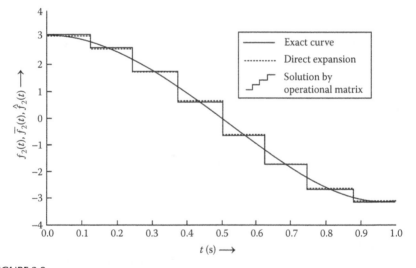

FIGURE 3.8

Differentiation of the function $f_1(t)$ using the operational matrix compared with the direct expansion of the result $f_2(t)$ in the Walsh domain for $m = 8$ and $T = 1$ second along with the exact curve.

3.2 Time Scaling of Operational Matrices

If λ is the scaling constant, the time scaling in time domain and Laplace domain are governed by the following well-known relations:

If, $f(t) \rightarrow f(\lambda t)$ for $t \rightarrow \lambda t$,

then

$$F(s) \rightarrow \frac{1}{\lambda} F\left(\frac{s}{\lambda}\right)$$

In the case of Walsh functions, the condition for orthonormality demands that in a scaled domain $t \to \lambda t$, we must have

$$\frac{1}{\lambda} \int_0^\lambda \phi_m(\lambda t)\phi_n(\lambda t)\mathrm{d}t = \delta_{mn} \tag{3.27}$$

where:
$\delta_{mn} = 1$ for $m = n$
$\delta_{mn} = 0$ for $m \neq n$.

In the above equation, $\phi_m(\lambda t)$ and $\phi_n(\lambda t)$ are $(m + 1)$th and $(n + 1)$th Walsh functions of the Walsh series in the scaled domain.

Hence, similar to $f(t)$, the time-scaling rule for Walsh function must be governed by

$$\phi_n(t) \to \phi_n(\lambda t) \quad \text{for} \quad t \to \lambda t \tag{3.28}$$

whereas Equation 3.27 formulates the rule for orthonormality.

If we denote the coefficients of the time-scaled Walsh series by $c_{n\lambda}$, then, following Equation 3.2, $c_{n\lambda}$ is given by

$$c_{n\lambda} = \frac{1}{\lambda} \int_0^\lambda \phi_n(\lambda t)f(t)\mathrm{d}t \tag{3.29}$$

3.2.1 Time-Scaled Operational Matrix for Integration

To derive the operational matrix for integration $\mathbf{G}_{\lambda(8)}$ in the scaled domain, we proceed in a similar manner as described in Section 3.1.1.

It should be noted that $\mathbf{G}_{\lambda(8)}$ deals with $\Phi_{(8)}(\lambda t)$ only. Hence, similar to Equation 3.1, we can write

$$\int \phi_0(\lambda t)\mathrm{d}t = f(t) \approx c_{0\lambda}\phi_0(\lambda t) + c_{1\lambda}\phi_1(\lambda t) + \cdots + c_{n\lambda}\phi_n(\lambda t) + \cdots \tag{3.30}$$

where:
$c_{0\lambda}, c_{1\lambda}, \ldots$ are the coefficients of the time-scaled Walsh series to be determined from Equation 3.29

Thus, integration of the first function of the time-scaled Walsh series yields

$$\int \phi_0(\lambda t)\mathrm{d}t = t \approx \frac{1}{\lambda}\left[\int_0^\lambda \phi_0(\lambda t)\, t\, \mathrm{d}t\right]\phi_0(\lambda t) + \frac{1}{\lambda}\left[\int_0^\lambda \phi_1(\lambda t)\, t\, \mathrm{d}t\right]\phi_1(\lambda t)$$
$$+ \frac{1}{\lambda}\left[\int_0^\lambda \phi_2(\lambda t)\, t\, \mathrm{d}t\right]\phi_2(\lambda t) + \cdots + \frac{1}{\lambda}\left[\int_0^\lambda \phi_n(\lambda t)\, t\, \mathrm{d}t\right]\phi_n(\lambda t) + \cdots$$

Considering only the first eight terms of the series, we have

$$\int \phi_0(\lambda t)dt \approx \frac{\lambda}{2}\phi_0(\lambda t) + \left(-\frac{\lambda}{4}\right)\phi_1(\lambda t) + \left(-\frac{\lambda}{8}\right)\phi_2(\lambda t) + \left(-\frac{\lambda}{16}\right)\phi_4(\lambda t) \quad (3.31)$$

In the compact matrix form, Equation 3.31 may be written as

$$\int \phi_0(\lambda t)dt \approx \begin{bmatrix} \frac{\lambda}{2} & -\frac{\lambda}{4} & -\frac{\lambda}{8} & 0 & -\frac{\lambda}{16} & 0 & 0 & 0 \end{bmatrix} \begin{bmatrix} \phi_0(\lambda t) \\ \phi_1(\lambda t) \\ \vdots \\ \phi_7(\lambda t) \end{bmatrix} \quad (3.32)$$

Similarly, if other time-scaled Walsh functions are integrated and expanded into time-scaled Walsh series, we can determine the related Walsh coefficients, and similar to Equation 3.6, we can obtain the following matrix relation:

$$\begin{bmatrix} \int \phi_0(\lambda t)dt \\ \int \phi_1(\lambda t)dt \\ \vdots \\ \int \phi_7(\lambda t)dt \end{bmatrix} \approx \begin{bmatrix} \frac{\lambda}{2} & -\frac{\lambda}{4} & -\frac{\lambda}{8} & 0 & -\frac{\lambda}{16} & 0 & 0 & 0 \\ \frac{\lambda}{4} & 0 & 0 & -\frac{\lambda}{8} & 0 & -\frac{\lambda}{16} & 0 & 0 \\ \frac{\lambda}{8} & 0 & 0 & 0 & 0 & 0 & -\frac{\lambda}{16} & 0 \\ 0 & \frac{\lambda}{8} & 0 & 0 & 0 & 0 & 0 & -\frac{\lambda}{16} \\ \frac{\lambda}{16} & 0 & 0 & 0 & 0 & 0 & 0 & 0 \\ 0 & \frac{\lambda}{16} & 0 & 0 & 0 & 0 & 0 & 0 \\ 0 & 0 & \frac{\lambda}{16} & 0 & 0 & 0 & 0 & 0 \\ 0 & 0 & 0 & \frac{\lambda}{16} & 0 & 0 & 0 & 0 \end{bmatrix} \begin{bmatrix} \phi_0(\lambda t) \\ \phi_1(\lambda t) \\ \vdots \\ \phi_7(\lambda t) \end{bmatrix} \quad (3.33)$$

or

$$\int \Phi_{(8)}(\lambda t)dt \approx \mathbf{G}_{\lambda(8)}\Phi_{(8)}(\lambda t) \quad (3.34)$$

Comparing Equations 3.6 and 3.7 with Equations 3.33 and 3.34, it is evident that

$$\mathbf{G}_{\lambda(8)} = \lambda\mathbf{G}_{(8)} \quad (3.35)$$

Hence, $\mathbf{G}_{\lambda(8)}$ is a modified operational matrix and could be termed as the operational matrix for integration in the scaled Walsh domain.

3.2.2 Time-Scaled Operational Matrix for Differentiation

To determine the modified operational matrix for differentiation in the scaled Walsh domain, we invert $\mathbf{G}_{\lambda(8)}$ to obtain $\mathbf{D}_{\lambda(8)}$. Hence,

$$\mathbf{D}_{\lambda(8)} = [\mathbf{G}_{\lambda(8)}]^{-1} = \frac{1}{\lambda}[\mathbf{G}_{(8)}]^{-1}$$

or   (3.36)

$$\mathbf{D}_{\lambda(8)} = \frac{1}{\lambda}\mathbf{D}_{(8)}$$

Obviously, if the first m Walsh functions of the series are considered, the generalized operational matrices in the scaled domain will be given by

$$\left.\begin{aligned} \mathbf{G}_{\lambda(m)} &= \lambda\mathbf{G}_{(m)} \\ \text{and} \\ \mathbf{D}_{\lambda(m)} &= \frac{1}{\lambda}\mathbf{D}_{(m)} \end{aligned}\right\}$$ (3.37)

The time-scaled operational matrices given by Equation 3.37 may be employed for the analysis of linear time-invariant SISO systems over different time ranges.

Now, we present the philosophy of the operational technique for system analysis.

3.3 Philosophy of the Proposed Walsh Domain Operational Technique

The input–output relationship of a linear time-invariant SISO system in Laplace domain is well known and is given by

$$C(s) = G(s)R(s)$$ (3.38)

where:
  $G(s)$ is the transfer function of the system
  $C(s)$ and $R(s)$ are the Laplace-transformed output $c(t)$ and input $r(t)$, respectively

The transfer function of a linear time-invariant system is the ratio of the Laplace transform of the output response function to the Laplace transform of the input driving function, under the assumption that all initial conditions are zero.

A linear time-invariant system may be defined by the following differential equation [9]:

$$a_0 c^{(n)} + a_1 c^{(n-1)} + \cdots + a_{n-1} c^{(1)} + a_n c = b_0 r^{(m)} + b_1 r^{(m-1)} + \cdots + b_{m-1} r^{(1)} + b_m r \quad (3.39)$$

where:

the superscripts (n), $(n-1)$,... and (m), $(m-1)$,... denote different orders of derivatives of the output c and input r of the system, respectively, and $n \geq m$

The transfer function of the system $G(s)$ is obtained by taking the Laplace transforms of both sides of Equation 3.39, under the assumption that all initial conditions are zero.

Thus,

$$G(s) = \frac{C(s)}{R(s)} = \frac{b_0 s^m + b_1 s^{m-1} + \cdots + b_{m-1} s + b_m}{a_0 s^n + a_1 s^{n-1} + \cdots + a_{n-1} s + a_n}$$

Since the operational matrix **D** performs in the Walsh domain like the Laplace operator s in the Laplace domain, we can convert $G(s)$ in Equation 3.38 to **WOTF** by simply replacing s by the operational matrix **D**. That is,

$$G(s)\big|_{s=D} = G(\mathbf{D}) = \mathbf{WOTF} \quad (3.40)$$

By the transformation of the above equation, the system transfer function is converted to an operational transfer function defined in the Walsh domain.

Since Walsh functions are defined in the time domain, representation of any time function $f(t)$ by a series combination of Walsh functions does not obscure the time variation of $f(t)$. Hence, our next step is to convert $R(s)$, or the corresponding time function $r(t)$, to Walsh domain and represent it by an input Walsh vector $\mathbf{R}\ \Phi(t)$.

At this point, we seek an equation having a form similar to that of Equation 3.38, which relates the input–output of a linear system in the Walsh domain instead of Laplace domain. To do this, we simply write

$$\left.\begin{array}{c} \mathbf{C}\ \Phi(t) = \mathbf{R}\ \mathbf{WOTF}\ \Phi(t) \\[1em] \text{or} \\[1em] \mathbf{C} = \mathbf{R}\ \mathbf{WOTF} \end{array}\right\} \quad (3.41)$$

Thus, $\mathbf{C}\ \Phi(t)$ in the above equation will actually give the output time response in the form of a series combination of Walsh functions.

The points to be noted about Equation 3.41 are as follows:

1. **WOTF** is always a square matrix having the same order as that of **D**, and it is operative upon Walsh functions only.
2. **R** and **C** are considered to be row matrices. If **R** and **C** are considered to be column vectors, then in Equation 3.41 these are to be replaced by their transposes.
3. The matrix multiplication on the right-hand side (RHS) of the equation is rather simple.
4. To determine the output response, no inverse transformation is necessary.

It is evident that the output response, obtained above in the sequency domain [10], gives the time response in a piecewise constant manner, that is, in the form of a staircase waveform. However, the method can accommodate any type of linear load once the differential equation governing the input–output relationship is known.

For nonlinear loads, however, a linearized model is necessary, and the time response can only be obtained for small-signal perturbation around a stable operating point.

The elegance of the method is enhanced due to the following facts:

1. It defines an operational transfer function in the sequency domain that is used for any kind of input stimulus, and a simple matrix multiplication yields the time response. The necessity of computing inverse transforms of the outputs for each kind of input is entirely avoided.
2. For square wave inputs, the Walsh functions enable us to express the input wave in sequency domain mostly by inspection only.
3. The method is well suited for computer manipulation.

In the Walsh domain analysis, since the output response is obtained as a staircase waveform, the method is more or less approximate. However, compared to exact solutions, the results obtained by Walsh method are in good agreement, particularly when 16 or more basis Walsh functions are used.

The **WOTF** is normally obtained from the conventional transfer function $G(s)$ of the system simply by replacing the Laplace operator s by the Walsh operational matrix for differentiation, **D.** However, if necessary, the **WOTF** can also be determined from system differential equations. In addition, the Walsh domain output waveform may be load current, load voltage, motor current, motor torque or speed, and so on, depending on the nature of the system transfer function $G(s)$ used.

Superiority of Equation 3.41 over Equation 3.38 may be judged from the following facts:

1. Determination of the inverse transform of $G(s)R(s)$ to obtain the output time response for a higher-order system with complex input waveform is very difficult and tedious.
2. Any input $r(t)$ may be expressed by a Walsh series $\mathbf{R}\,\Phi(t)$. In particular, pulse trains, square wave functions, and inputs of similar nature could be expressed in the Walsh domain very easily. The same task is somewhat difficult in Laplace domain, where the inverter and chopper waveforms cannot easily be expressed.
3. $R\,\Phi(t)$ is always a row/column matrix, enabling the matrix multiplication on the RHS of Equation 3.41 rather simple.
4. Only the computation of **WOTF** needs the help of a computer. But once computed, the same system, having the same **WOTF**, can be tested for different input waveforms without additional labor. In Laplace domain analysis, for each type of input waveform, the inverse transform of the product $[G(s)R(s)]$ is to be computed separately.

The only disadvantage of Equation 3.41 is that the time response obtained in the form of $\mathbf{C}\,\Phi(t)$ is somewhat approximate because of its inherent stepped nature. However, the error may be reduced by increasing the number of basis Walsh functions used during waveform synthesis.

It should be noted that if we use the operational matrix $\mathbf{G}$ instead of $\mathbf{D}$ to compute the **WOTF** of a particular system, the same **WOTF** is obtained. Only, to compute the **WOTF**, first we have to convert $G(s)$ to $G_1(1/s)$ and then substitute the operational matrix $\mathbf{G}$ in place of $1/s$ in $G_1(1/s)$.

At this point, we seek the help of block pulse functions shown in Figure 2.8. Block pulse functions are related to Walsh functions [8] by relation (2.16).

$$\Phi_{(m)} = \mathbf{W}_{(m)}\Psi_{(m)}$$

where:
$\mathbf{W}_{(m)}$ is the Walsh matrix

For $m = 8$, the Walsh matrix has the following form [10]:

$$\mathbf{W}_{(8)} = \begin{bmatrix} 1 & 1 & 1 & 1 & 1 & 1 & 1 & 1 \\ 1 & 1 & 1 & 1 & -1 & -1 & -1 & -1 \\ 1 & 1 & -1 & -1 & 1 & 1 & -1 & -1 \\ 1 & 1 & -1 & -1 & -1 & -1 & 1 & 1 \\ 1 & -1 & 1 & -1 & 1 & -1 & 1 & -1 \\ 1 & -1 & 1 & -1 & -1 & 1 & -1 & 1 \\ 1 & -1 & -1 & 1 & 1 & -1 & -1 & 1 \\ 1 & -1 & -1 & 1 & -1 & 1 & 1 & -1 \end{bmatrix} \tag{3.42}$$

The set of block pulse functions is more fundamental than the Walsh functions, and hence, construction of any function in the time domain is easier with block pulse functions than with Walsh functions.

In Equation 3.42, each row of Walsh matrix represents a Walsh function expressed via a row of block pulse functions. Thus, the sum of eight block pulse functions (i.e., eight elements of any row of the Walsh matrix) gives one Walsh function in this case.

Thus, if we consider $\mathbf{R'}$ to be the block pulse coefficient matrix of input $r(t)$ and $\mathbf{C'}$ to be block pulse coefficient matrix of output $c(t)$, $\mathbf{R'}\,\Psi(t)$ could be formed from $r(t)$ merely by inspection. To convert $\mathbf{R'}\,\Psi$ to $\mathbf{R}\,\Phi$, we use the transformation represented by Equations 2.16 and 2.18.

Thus,

$$\mathbf{R'}\Psi = \mathbf{R'}\,\mathbf{W}^{-1}\Phi = \frac{1}{m}\,\mathbf{R'}\,\mathbf{W}\,\Phi \triangleq \mathbf{R}\Phi \tag{3.43}$$

where use is made of relation (2.18).

From Equation 3.41,

$$\mathbf{C}\Phi = \mathbf{R}\,\mathbf{WOTF}\,\Phi = \frac{1}{m}\,\mathbf{R'}\,\mathbf{W}\,\{\mathbf{WOTF}\}\,\Phi \quad [\text{using relation (3.43)}]$$

$$\mathbf{C}\mathbf{W}\,\Psi = \frac{1}{m}\,\mathbf{R'}\,\mathbf{W}\,\{\mathbf{WOTF}\}\,\mathbf{W}\,\Psi \quad [\text{using relation (2.16)}]$$

Using $\mathbf{C}\mathbf{W} \triangleq \mathbf{C'}$, we have

$$\mathbf{C'} = \mathbf{R'}\left[\frac{1}{m}\mathbf{W}\,\{\mathbf{WOTF}\}\mathbf{W}\right] \tag{3.44}$$

So the matrix $[(1/m)\mathbf{W}\,\{\mathbf{WOTF}\}\mathbf{W}]$ is the operational transfer function in BPF domain that is termed block pulse operational transfer function (**BPOTF**) because it relates the block pulse vectors $\mathbf{C'}$ and $\mathbf{R'}$, and it is operative upon the Ψ vector only.

Since easier reconstruction of output response $c(t)$ is possible if block pulse functions are used, we use the transformation of Equation 3.44 as and when necessary. However, the operational technique, whether we use Walsh functions or block pulse functions, produces the same result.

3.4 Analysis of a First-Order System with Step Input

Let us consider a first-order system having a transfer function given by

$$G_1(s) = \frac{1}{(s+1)} \tag{3.45}$$

We apply a step input to the system and study the time response $c(t)$. Suppose we are interested in the solution for $0 \leq t \leq 4$ seconds.

Since Walsh functions are defined in the interval of 0 to 1 second, the scaling factor λ in this case would be 4. Consequently, the operational matrix that would replace s in $G_1(s)$ above would be $\mathbf{D}_\lambda = \mathbf{D}/4$ as per Equation 3.37.

Hence, the **WOTF** for the first-order system is

$$\mathbf{WOTF1} = \left[\frac{\mathbf{D}}{4} + \mathbf{I} \right]^{-1} \tag{3.46}$$

In the above equation, if we consider the first eight terms of the Walsh series, the operational matrix $\mathbf{D}$ is of order 8 and $\mathbf{I}$ is an identity matrix of the same order. It is evident that after substitution of $\mathbf{D}$ in Equation 3.46 from Equation 3.19, **WOTF1** will turn out to be a square matrix of order 8.

Computation of the RHS of Equation 3.46 yields **WOTF1** as

WOTF1 =

$$\begin{bmatrix}
0.75419 & -0.18939 & -0.11567 & -0.08913 & -0.06145 & -0.04735 & -0.02892 & -0.02228 \\
0.18939 & 0.37540 & 0.08913 & -0.29392 & 0.04735 & -0.15614 & 0.02228 & -0.07348 \\
0.11567 & 0.08913 & 0.11325 & 0.04194 & 0.02892 & 0.02228 & -0.22168 & 0.01048 \\
-0.89128 & 0.29392 & -0.04194 & 0.19714 & -0.02228 & 0.07348 & -0.01048 & -0.20071 \\
0.61450 & 0.04735 & 0.02892 & 0.02228 & 0.01536 & 0.01184 & 0.00723 & 0.00557 \\
-0.47349 & 0.15614 & -0.02228 & 0.07348 & -0.01184 & 0.03904 & -0.00557 & 0.01837 \\
-0.28917 & -0.02228 & 0.22168 & -0.01048 & -0.00723 & -0.00557 & 0.05542 & -0.00262 \\
0.22282 & -0.07348 & 0.01048 & 0.20071 & 0.00557 & -0.01837 & 0.00262 & 0.05018
\end{bmatrix} \tag{3.47}$$

The input signal to the system is a unit step function, which is nothing but $\phi_0(t)$, the first member of the Walsh function family. Hence, $u(t)$ can be expressed as

$$u(t) = \mathbf{R}\,\Phi_{(8)} = [1 \quad 0 \quad 0 \quad 0 \quad 0 \quad 0 \quad 0 \quad 0]\Phi_{(8)} \tag{3.48}$$

where:
    $\Phi_{(8)}$ is a column vector of Walsh functions having its components as ϕ_0, $\phi_1, \ldots, \phi_7$

In terms of block pulse functions, $u(t)$ can equivalently be represented by the following relation:

$$u(t) = \mathbf{R}'\,\Psi_{(8)} = [1 \quad 1 \quad 1 \quad 1 \quad 1 \quad 1 \quad 1 \quad 1]\,\Psi_{(8)}$$

where:

$\Psi_{(8)}$ is a column vector of block pulse functions having its components as $\psi_0, \psi_1, \ldots, \psi_7$.

Following Equation 3.44

$$C' = R' \left[\frac{1}{m} W \{WOTF\} W \right] \tag{3.49}$$

where:

W is also of order 8 and is given by Equation 3.42

$$C' \Psi_{(8)} = [1 \quad 1 \quad 1 \quad 1 \quad 1 \quad 1 \quad 1 \quad 1] \frac{1}{8} W \{WOTF1\} W \ \Psi_{(8)} \tag{3.50}$$

Substituting **W** and **WOTF1** from Equations 3.42 and 3.47, respectively, computation of the matrix product on the RHS of Equation 3.50 yields the output time response as

$$C' \Psi_{(8)} = [0.200 \quad 0.520 \quad 0.712 \quad 0.827 \quad 0.896 \quad 0.938 \quad 0.963 \quad 0.978] \ \Psi_{(8)} \tag{3.51}$$

The Walsh/block pulse response is shown in Figure 3.9 (Program B.9 in Appendix B) along with the actual time response of the first-order system.

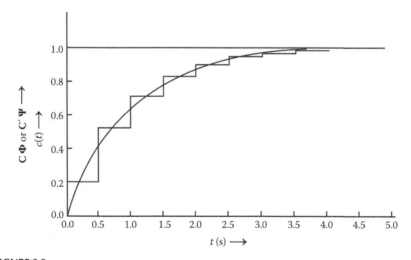

FIGURE 3.9

Time response of a first-order system to a step input and its equivalent Walsh/block pulse representation for $m = 8$ and $T = 4$ seconds.

3.5 Analysis of a Second-Order System with Step Input

We choose a second-order system having a transfer function:

$$G_2(s) = \frac{1}{s^2 + 0.6s + 1} \tag{3.52}$$

For simplicity of computation, the natural frequency of the system is chosen as 1 rad/second. However, the damping ratio is taken to be 0.3 so that the system is underdamped and shows pronounced oscillations before the response settles down.

To obtain a solution for a period $0 \leq t \leq 8$ seconds due to the application of a step input, we use Equation 3.37 and substitute the operational matrix $\mathbf{D}_\lambda = \mathbf{D}/8$ [for simplicity, we drop the dimension indicator (m)] in place of s to obtain the **WOTF**. Hence,

$$\mathbf{WOTF2} = \left[\frac{\mathbf{D}^2}{64} + \frac{0.6\mathbf{D}}{8} + \mathbf{I} \right]^{-1} \tag{3.53}$$

where:
  $\mathbf{I}$ is a unit matrix

Now, for $m = 8$, $\mathbf{D}$ is substituted from Equation 3.19 into the above expression and the computation of the RHS of Equation 3.53 yields the **WOTF** as

WOTF2 =

$$\begin{bmatrix}
0.91877 & -0.08182 & -0.14540 & -0.29092 & -0.05513 & -0.11836 & -0.00401 & -0.06182 \\
0.08182 & 0.75513 & 0.29092 & -0.72725 & 0.11836 & -0.29186 & 0.06182 & -0.12765 \\
0.14540 & 0.29092 & -0.24470 & 0.20900 & 0.00401 & 0.06182 & -0.19482 & -0.11034 \\
-0.29092 & 0.72725 & -0.20900 & 0.17329 & -0.06182 & 0.12765 & 0.11034 & -0.41551 \\
0.05513 & 0.11836 & 0.00401 & 0.06182 & -0.00377 & 0.01506 & -0.03515 & -0.05418 \\
-0.11836 & 0.29186 & -0.06182 & 0.12765 & -0.01506 & 0.02634 & 0.05418 & -0.14351 \\
-0.00401 & -0.06182 & 0.19482 & 0.11034 & 0.03515 & 0.05418 & -0.25273 & 0.08535 \\
0.06182 & -0.12765 & -0.11034 & 0.41551 & -0.05418 & 0.14351 & -0.08535 & -0.08202
\end{bmatrix} \tag{3.54}$$

Following the procedure outlined in the previous section, the output response in this case is given by

$$\mathbf{C}' \Psi_{(8)} = [1 \quad 1 \quad 1 \quad 1 \quad 1 \quad 1 \quad 1 \quad 1] \frac{1}{8} \mathbf{W} \{\mathbf{WOTF2}\} \mathbf{W} \ \Psi_{(8)} \tag{3.55}$$

Substitution of **W** and **WOTF2** from Equations 3.42 and 3.54, respectively, and the subsequent computation of the matrix product on the RHS of Equation 3.55 gives the output as

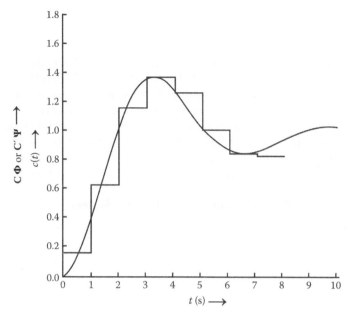

FIGURE 3.10
Time response of a second-order system to a step input and its equivalent Walsh/block pulse representation for $m = 8$ and $T = 8$ seconds.

$$\mathbf{C'} \, \Psi_{(8)} = [0.161 \quad 0.640 \quad 1.166 \quad 1.381 \quad 1.267 \quad 1.025 \quad 0.860 \quad 0.850] \, \Psi_{(8)} \quad (3.56)$$

The output Walsh/block pulse response is shown in Figure 3.10 with the actual time response of the second-order system (Program B.10 in Appendix B).

3.6 Oscillatory Phenomenon in Walsh Domain System Analysis

Relating to approximation of a particular function by partial sums of a Fourier series, the Gibbs phenomenon is a well-known feature [9]. If the function $f(t)$ to be approximated has discontinuities, the partial sum approximating the function at such discontinuities shows a noticeable oscillation with a maximum overshoot of 9%. If the number of terms in the Fourier series, used for approximating $f(t)$, is increased to infinity, even there is no reduction to the figure of 9% mentioned above.

A similar phenomenon has been observed by J. L. Walsh [12] when he used a new closed set of orthonormal functions, later named Walsh functions, for approximating a function $f(t)$ at discontinuities. Supposing that $f(t)$ has a discontinuity at $t = t_1$, if the point t_1 is dyadically irrational, $f(t)$ cannot be expanded in terms of Walsh functions, and there occurs a phenomenon quite analogous to Gibbs phenomenon for Fourier series. Walsh, in his derivation, used a function

$$f(t_1) = \begin{Bmatrix} 1, & 0 \le t < t_1 \\ 0, & t_1 < t \le 1 \end{Bmatrix}$$

to analyze its nonconvergence at $t = t_1$ when t_1 was dyadically irrational.

In Ref. [12], it was claimed that Walsh domain analysis handles functions with discontinuities more elegantly than Fourier analysis due to the fact that there is no oscillation like Gibbs phenomenon. However, our results are contrary to this claim, and this section establishes a new condition that guides the occurrence of such oscillations for Walsh domain solution of first-order systems with unit step input. Representative curves and tables are also presented to show the nature of such oscillations, and few relevant observations are made.

3.6.1 Oscillatory Phenomenon in a First-Order System

Consider a first-order linear time-invariant SISO system having a Laplace domain transfer function given by

$$G(s) = \frac{1}{s + a} \tag{3.57}$$

where:
1/a is the time constant of the system

The **WOTF** for the same system corresponding to $G(s)$, say **WOTF3**, is constructed according to the method expanded in Section 3.3. Therefore, the **WOTF** can be written as

$$\textbf{WOTF3} = [\textbf{D} + a\textbf{I}]^{-1} \tag{3.58}$$

where:
I is a unit matrix of the same order as that of **D**

We consider the first m Walsh functions of the series, and if we are interested in the solution over a region of $0 \le t \le \lambda$ seconds, according to Equation 3.37, **WOTF** becomes

$$\textbf{WOTF3} = \left[\frac{\textbf{D}_{(m)}}{\lambda} + a\textbf{I}_{(m)} \right]^{-1} \tag{3.59}$$

Following the procedure outlined in Section 3.3, the response of such a system due to a unit step function input is given by

$$\textbf{C}\Phi_{(m)} = \begin{bmatrix} 1 & \underbrace{0 \quad 0 \quad \dots \quad 0}_{(m-1) \text{ zeros}} \end{bmatrix} \left[\frac{\textbf{D}_{(m)}}{\lambda} + a\textbf{I}_{(m)} \right]^{-1} \Phi_{(m)} \tag{3.60}$$

The RHS of the above equation is evaluated for $\lambda = 4$ and for different values of m and a. Since m is of the form 2^p, three values of m—namely, 4, 8, and 16—were

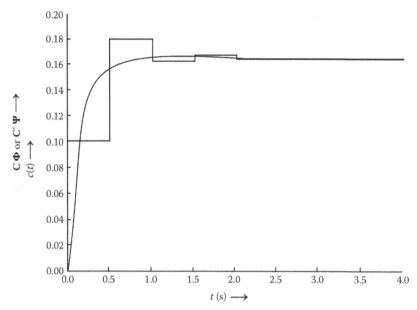

FIGURE 3.11

Comparison between exact solution $c(t)$ and Walsh domain solution $\mathbf{C}\,\Phi(t)$ for the first-order system for $m = 8$, $\lambda = 4$, and $a = 6$, respectively. (Reprinted from Deb, A. and Fountain, D. W., *IEEE Trans. Circuits Syst.*, 38, 945–948, 1991. With permission.)

considered and by varying the value of a, the output responses for different combinations of λ, m, and a were obtained. It was found from the computed results that the output response shows distinct oscillations like the Gibbs phenomenon for different combinations of λ, m, and a. It was found from the computed results that the output response shows distinct oscillations for

$$a\lambda > 2m \tag{3.61}$$

and no oscillations were observed for

$$a\lambda \le 2m \tag{3.62}$$

This is illustrated by time-domain solutions obtained from Walsh analysis and the actual solution, shown in Figure 3.11 (Program B.11 in Appendix B), for $m = 8$, $\lambda = 4$, and $a = 6$. It may be noted that the actual solution for a first-order system with transfer function given by Equation 3.57 and a unit step is

$$c(t) = a^{-1}[1 - \exp(-at)], \quad t \ge 0 \tag{3.63}$$

3.6.2 Analytical Study of the Oscillatory Phenomenon

We start with a standard decomposition of the operational matrix for differentiation $\mathbf{D}$ based on the "discrete Walsh matrix" $\mathbf{W}$ [3,13–14].

Using **W** as a transformation matrix, it can be shown that

$$\mathbf{D}_{(m)} = \frac{1}{m}\mathbf{W}_{(m)}\,\mathbf{D}'_{(m)}\mathbf{W}_{(m)} \tag{3.64}$$

where:
D$'_{(m)}$ is given by the upper triangular Toeplitz form:

$$\mathbf{D}'_{(m)} = 4m\begin{bmatrix} \frac{1}{2} & -1 & 1 & \cdots & -1 \\ 0 & \frac{1}{2} & -1 & \cdots & 1 \\ \vdots & \vdots & \vdots & \vdots & \vdots \\ 0 & 0 & 0 & \cdots & \frac{1}{2} \end{bmatrix} \tag{3.65}$$

To apply this transformation for the present case, we make the following approximations: the input $r(t) \approx \mathbf{R}\boldsymbol{\Phi}_{(m)}$ and the output $c(t) \approx \mathbf{C}\boldsymbol{\Phi}_{(m)}$, where **R** and **C** are Walsh vectors.

Hence,

$$\mathbf{R} = \begin{bmatrix} 1 & \underbrace{0 \quad 0 \quad \cdots \quad 0}_{(m-1)\ \text{zeros}} \end{bmatrix}, \text{ since } r(t) = u(t) \tag{3.66}$$

Then, according to Equation 3.59, we need to solve the system:

$$\mathbf{C}\left[\frac{\mathbf{D}_{(m)}}{\lambda} + a\mathbf{I}_{(m)}\right] = \mathbf{R} \tag{3.67}$$

for the unknown $(1 \times m)$ row vector **C**. We shall use the $\mathbf{W}_{(m)}$ transformation to reduce this equation to a simpler form. Using Equation 3.64 to substitute for $\mathbf{D}_{(m)}$ gives

$$\mathbf{C}\left[\frac{1}{m}\mathbf{W}_{(m)}\,\mathbf{D}'_{(m)}\mathbf{W}_{(m)}\left(\frac{1}{\lambda}\right) + a\mathbf{I}_{(m)}\right] = \mathbf{R} \tag{3.68}$$

Multiplying both sides by $\mathbf{W}_{(m)}$ and using Equation 2.18, we have

$$\mathbf{C}\mathbf{W}_{(m)}\left[\frac{\mathbf{D}'_{(m)}}{\lambda} + a\mathbf{I}_{(m)}\right] = \mathbf{R}\mathbf{W}_{(m)} \tag{3.69}$$

Now we define

$$\mathbf{C}' \equiv \mathbf{C}\mathbf{W}_{(m)} = [c'_0\ c'_1\ \dots\ c'_{m-1}] \tag{3.70}$$

It may be noted that [8]

$$\mathbf{RW}_{(m)} = \begin{bmatrix} 1 & \underbrace{1 \quad 1 \quad \cdots \quad 1}_{m \text{ columns}} \end{bmatrix}, \text{ for the unit step input}$$

Here, the elements of the row vector are the block pulse spectral coefficients, since the block pulse functions are related to Walsh functions by the relation $\Phi_{(m)} = \mathbf{W}_{(m)} \Psi_{(m)}$ [15]. Obviously, $\mathbf{D}'_{(m)}$ is the operational matrix for differentiation in the block pulse domain. Hence the values c'_k are the actual values of the block pulse spectral coefficients, giving the block pulse approximation of the output in $(k+1)$th subinterval [8,13]. So, in the interval $k(\lambda/m) \le t \le (k+1)(\lambda/m)$, we have

$$\mathbf{C}\Phi_{(m)} = \mathbf{C}'\Psi_{(m)} = \sum_{k=0}^{m-1} c'_k \psi_k$$

Thus, in what follows, we concentrate on the values of c'_k. Substituting Equation 3.65 in Equation 3.69 and using the notation of Equation 3.70, we can write out this system of equation explicitly as

$$c'_0 \left(\frac{2m}{\lambda} + a \right) = 1$$

$$c'_1 \left(\frac{2m}{\lambda} + a \right) = \left(\frac{4m}{\lambda} \right) c'_0 + 1$$

$$c'_2 \left(\frac{2m}{\lambda} + a \right) = -\left(\frac{4m}{\lambda} \right) c'_0 + \left(\frac{4m}{\lambda} \right) c'_1 + 1$$

$$c'_k \left(\frac{2m}{\lambda} + a \right) = -\left(\frac{2m}{\lambda} + a \right) c'_{k-1} + \left(\frac{4m}{\lambda} \right) c'_{k-1} + 2$$

$$= \left(\frac{2m}{\lambda} - a \right) c'_{k-1} + 2 \quad k = 0, 1, \ldots, m-1$$

Hence, the problem reduces to analyzing the set of equations:

$$c'_0 = \frac{\lambda}{2m + a\lambda}$$

$$c'_k = \frac{2m - a\lambda}{2m + a\lambda} c'_{k-1} + \left(\frac{2\lambda}{2m + a\lambda} \right) \equiv Ac'_{k-1} + \left(\frac{2\lambda}{2m + a\lambda} \right) \qquad (3.71)$$

Solving the system, we obtain the general form:

$$c'_k = \frac{1}{a} \left[1 - A^k \left(\frac{2m}{2m + a\lambda} \right) \right] \quad k = 0, 1, \ldots, m-1. \qquad (3.72)$$

The above equation will now be studied to bring out all possible interpretations of the oscillatory behavior.

Case 1: $a\lambda = 2m$

When $k = 0$, from Equation 3.71, we have $c_0' = \lambda/2m + a\lambda$. But since $a\lambda = 2m$, $c_0' = 1/2a$. Also, for this case, $A = 0$. Substituting in Equation 3.72, we have $c_1' = c_2' = \cdots = c_{m-1}' = (1/a)$. Thus, there is no oscillation and the spectral coefficients approximate the output waveform quite faithfully.

Case 2: $a\lambda < 2m$

For this case, A is positive but less than unity. So the term A^k in Equation 3.72 will decrease with increasing k. This will give $c_0' < c_1' < \cdots < c_{m-1}'$. When m becomes very large, the second term within the square bracket of Equation 3.72 will tend to zero, and the spectral coefficients will converge to the expected value of $1/a$. Thus, no oscillation will occur.

Case 3: $a\lambda > 2m$

For this case, denote $B = -A$. Then, Equation 3.72 can be written as

$$c_k' = \frac{1}{a}\left[1 - (-1)^k B^k\left(\frac{2m}{2m + a\lambda}\right)\right] \quad k = 0,1,\ldots,m-1 \qquad (3.73)$$

Since B is positive, the sign of the second term within the square bracket of Equation 3.73 is totally dependent on the factor $(-1)^k$. Hence, when k is odd, the sign of the second term within the square bracket of Equation 3.73 becomes negative, and the resulting spectral component c_k', which is of even order (since the subscript k starts from 0), becomes greater than $1/a$.

Similarly, when k is even, the sign of the second term within the square bracket of Equation 3.73 becomes positive, and the resulting spectral component c_k', which is of odd order, becomes less than $1/a$. This results in the oscillations indicated earlier.

Since A is less than unity, it is evident that as k increases, the value of the second term within the square bracket decreases and the oscillations become less noticeable. Thus, the overshoot also decreases. Hence, from Equation 3.73, it is clear that the overshoot will be maximum for the minimum odd value of k, which is unity. Substituting $k = 1$ and the value of B in Equation 3.73, the corresponding spectral coefficient c_1' is given by

$$c_1' = \frac{1}{a}\left[1 + \frac{2m(a\lambda - 2m)}{(2m + a\lambda)^2}\right]$$

So the maximum percentage overshoot M is

$$M \times 100\% = \frac{2m(a\lambda - 2m)}{(2m + a\lambda)^2} \times 100\%$$

To find out when this maximum percentage overshoot will be the largest, we express M as

$$M = \frac{(p-1)}{(p+1)^2} \tag{3.74}$$

where:
$$p = \frac{a\lambda}{2m}$$

Now differentiating M with respect to p and equating to 0, we have $p = 3$, that is, $a\lambda = 6m$. This gives a value of $M = 0.125$ as shown in Figure 3.16. For $\lambda = 4$ and $m = 16$, the value of a is 24.

Attention is now drawn to the striking similarity between Equations 3.63 and 3.72. As per the standard theory, the Walsh or block pulse spectral coefficient of any real valued square-integrable function of Lebesgue measure $f(t)$ in any subinterval tries to equal the average value of the function in that particular interval. In the $(k + 1)$th subinterval $k(\lambda/m) \leq t \leq (k+1)(\lambda/m)$, c'_k from Equations 3.71 and 3.74 is given by

$$c'_k = \frac{1}{a}\left[1 - \frac{(1-p)^k}{(1+p)^{k+1}}\right] \tag{3.75}$$

From Equation 3.63, the average value of $c(t)$ in the same subinterval is given by

$$\bar{c}'_k = \frac{1}{\left(\dfrac{\lambda}{m}\right)} \frac{1}{a} \int_{k\frac{\lambda}{m}}^{(k+1)\frac{\lambda}{m}} [1-\exp(-at)]\, dt$$

$$= \frac{1}{a}\left[1 - \frac{1}{2p}[\exp(-2pk) - \exp\{-2p(k+1)\}]\right] \tag{3.76}$$

Comparison of Equations 3.75 and 3.76 reveals certain interesting facts:

1. For all possible values of p (remembering that p can never be negative), the term within the inner square bracket of Equation 3.76 can never be negative.

2. For $p > 1$ and when k is odd, the second term within the square bracket of Equation 3.75 becomes negative, giving rise to an overshoot in the system output, that is, $c'_k > 1/a$.

3. Thus, while Equation 3.76 is completely stable, Equation 3.75 is not. But for allowable values of p (i.e., $p \leq 1$), the results of the two equations match very closely. This is indicated in Table 3.5, which lists the difference $a(c'_k - \bar{c}'_k)$ for several values of p and k.

TABLE 3.5

Typical Values of $a(c_k^l - \bar{c}_k^l)$ for $p \leq 1$

$a(c_k^l - \bar{c}_k^l)$	For $k = 0$	For $k = 3$	For $k = 6$	For $k = 9$
$p = 0.02$	-1.281×10^{-4}	-9.973×10^{-5}	-7.611×10^{-5}	-5.656×10^{-5}
$p = 0.25$	-1.306×10^{-2}	2.790×10^{-3}	1.855×10^{-3}	6.799×10^{-4}
$p = 0.60$	-4.266×10^{-2}	6.146×10^{-3}	2.822×10^{-4}	9.495×10^{-6}
$p = 0.95$	-6.523×10^{-2}	1.489×10^{-3}	5.011×10^{-6}	1.677×10^{-8}

This oscillatory phenomenon may seem similar to the well-known Gibbs phenomenon, but it has properties quite unlike the Gibbs phenomenon. Hence while analyzing the system in the Walsh domain, care should be taken to follow the constraints of Equations 3.61 and 3.62 so that undesired oscillations do not affect the system output.

Maximum percentage overshoots for different situations are computed and tabulated in Tables 3.6 through 3.8. Table 3.9 summarizes the variations of maximum percentage overshoot with a for $m = 16$ and $\lambda = 4$, where a varies from 8 to 11 considering very small steps.

Based on these tables, Figures 3.12 through 3.16 (Program B.12 in Appendix B) are drawn to show the respective variations graphically. It may be pointed out that the output response of the first-order system considered in the analysis has no discontinuity, but still oscillations were observed in Walsh analysis and the same were found to be guided by the constraints in Equations 3.61 and 3.62. Further, the reasons for such oscillations were investigated analytically in detail and special cases were studied.

From Figures 3.14 through 3.16, it could be observed that the case for $m = 16$ and $\lambda = 4$ was studied in more detail. Figure 3.14 shows the characteristics for a slow variation from 8 to 11, whereas Figure 3.15 shows the variation of maximum percentage overshoot with ranges of a from 0 to 100 and 100 to 1000 for $m = 16$ and $\lambda = 4$. Figure 3.16 shows the variation of maximum percentage overshoot with whole range of a in logarithmic scale, and it could

TABLE 3.6

Variation of Maximum Percentage Overshoot with a Varied with Steps of 0.01 for $m = 4$ and $\lambda = 4$

Reciprocal of System Time Constant (a)	Maximum Overshoot (%)	Reciprocal of System Time Constant (a)	Maximum Overshoot (%)
2.00	0.000000	2.06	0.729250
2.01	0.124621	2.07	0.844563
2.02	0.248465	2.08	0.961120
2.03	0.369461	2.09	1.074490
2.04	0.489606	2.10	1.188601
2.05	0.608856	–	–

TABLE 3.7

Variations of Maximum Percentage Overshoot with a Varied with Steps of 0.1 for $m = 8$ and $\lambda = 4$

Reciprocal of System Time Constant (a)	Maximum Overshoot (%)	Reciprocal of System Time Constant (a)	Maximum Overshoot (%)
4.0	0.0000	5.2	5.6692
4.1	0.6058	5.3	6.0106
4.2	1.1864	5.4	6.3368
4.3	1.7380	5.5	6.6450
4.4	2.2648	5.6	6.9432
4.5	2.7665	5.7	7.2227
4.6	3.2424	5.8	7.4914
4.7	3.6961	5.9	7.7517
4.8	4.1312	6.0	8.0000
4.9	4.5415	6.1	8.2323
5.0	4.9350	6.2	8.4566
5.1	5.3099	6.3	8.6687

TABLE 3.8

Variations of Maximum Percentage Overshoot with a Varied from 1 to 10^8 with $m = 16$ and $\lambda = 4$

Reciprocal of System Time Constant (a)	Maximum Overshoot (%)	Reciprocal of System Time Constant (a)	Maximum Overshoot (%)
1	0.000	100	6.310
10	4.930	120	5.468
15	10.587	150	4.551
16	11.110	180	3.893
17	11.519	210	3.400
18	11.835	240	3.018
19	12.070	270	2.711
20	12.244	300	2.462
21	12.365	330	2.254
22	12.445	360	2.078
23	12.487	390	1.927
24	12.502	420	1.795
25	12.487	500	1.525
35	11.683	1000	0.780
45	10.535	2000	0.394
55	9.472	10^4	0.070
65	8.557	10^5	0.000
75	7.7750	10^6	0.000
85	7.1200	10^7	0.000
95	6.5520	10^8	0.000

TABLE 3.9

Variations of Maximum Percentage Overshoot with a between 8 and 11 Varied with Steps of 0.1 for $m = 16$ and $\lambda = 4$

Reciprocal of System Time Constant (a)	Maximum Overshoot (%)
8.0	0
8.2	0.610
8.4	1.190
8.6	1.742
8.8	2.267
9.0	2.768
9.2	3.245
9.4	3.699
9.6	4.132
9.8	4.545
10.0	4.938
10.2	5.313
10.4	5.671
10.6	6.012
10.8	6.338
11.0	6.648

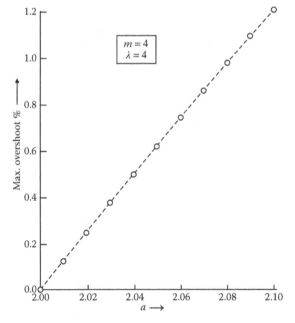

FIGURE 3.12

Variation of maximum percentage overshoot with a varied with steps of 0.01 for $m = 4$ and $\lambda = 4$. (Reprinted from Deb, A. and Fountain, D. W., *IEEE Trans. Circuits Syst.*, 38, 945–948, 1991. With permission.)

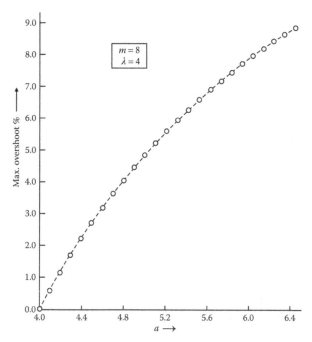

FIGURE 3.13
Variation of maximum percentage overshoot with *a* varied with steps of 0.1 for *m* = 8 and *λ* = 4. (Reprinted from Deb, A. and Fountain, D. W., *IEEE Trans. Circuits Syst.*, 38, 945–948, 1991. With permission.)

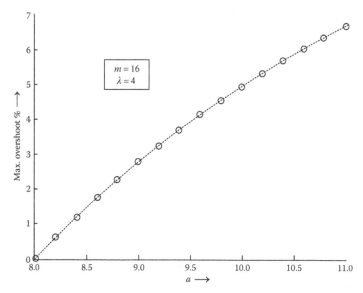

FIGURE 3.14
Variation of maximum percentage overshoot with *a* varied with steps of 0.2 for *m* = 16 and *λ* = 4.

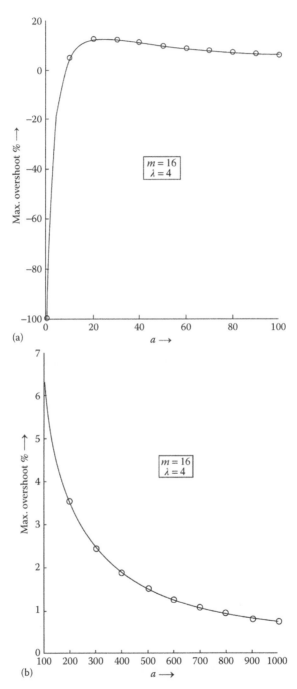

FIGURE 3.15
Variation of maximum percentage overshoot with ranges of a from (a) 0 to 100 and (b) 100 to 1000 (for $m = 16$ and $\lambda = 4$).

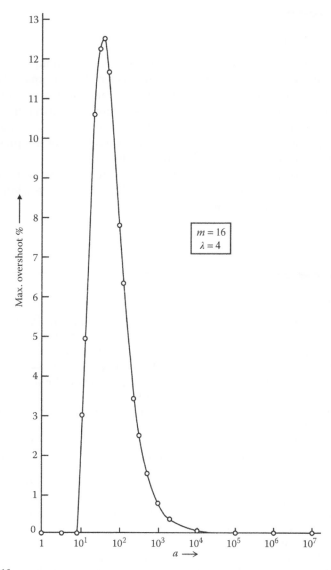

FIGURE 3.16
Variation of maximum percentage overshoot with whole range of *a* in logarithmic scale for *m* = 6 and *λ* = 4. (Reprinted from Deb, A. and Fountain, D. W., *IEEE Trans. Circuits Syst.*, 38, 945–948, 1991. With permission.)

be seen that the overshoot is a maximum of 12.5% for *a* = 24 as expected. When *a* is so large that $s \ll a$ in Equation 3.57, the response becomes a unit step function like the applied input. Since Walsh functions could exactly approximate a unit step function, there is no oscillation as is seen from Figure 3.16.

3.7 Conclusion

In this chapter, we have introduced a new operational technique for analyzing linear time-invariant SISO systems in the scaled Walsh domain. From the results, we see that the Walsh domain solutions are reasonably close to the exact solutions. It is found that there is a restriction on Walsh domain solution guided by Equations 3.61 and 3.62. However, by choosing the scaling constant λ properly, we can always avoid the possibility of the presence of undesired oscillations in the Walsh domain solution.

We now proceed to analyze pulse-fed SISO systems in the Walsh domain and then apply the presented operational technique to pulse-width modulated power electronic systems.

References

1. Deb, A. and Datta, A. K., Analysis of pulse-fed power electronic circuits using Walsh function, *Int. J. Electron.*, Vol. **62**, No. 3, pp. 449–459, 1987.
2. Corrington, M. S., Solution of differential and integral equations with Walsh functions, *IEEE Trans. Circuit Theory*, Vol. **CT-20**, No. 5, pp. 470–476, 1973.
3. Chen, C. F. and Hsiao, C. H., A state space approach to Walsh series solution of linear systems., *Int. J. Syst. Sci.*, Vol. **6**, No. 9, pp. 833–858, 1975.
4. Rao, G. P. and Sivakumar, L., System identification via Walsh functions, *Proc. IEE*, Vol. **122**, No. 10, pp. 1160–1161, 1975.
5. Le Van, T., Tam, L. C. and Van Houtte, N., On direct algebraic solution of linear differential equations using Walsh transforms, *IEEE Trans. Circuits Syst.*, Vol. **22**, No. 5, pp. 419–422, 1975.
6. Chen, C. F. and Hsiao, C. H., Time domain synthesis via Walsh functions, *Proc. IEE*, Vol. **122**, No. 5, pp. 565–570, 1975.
7. Chen, C. F. and Hsiao, C. H., Design of piecewise constant gains for optimal control via Walsh functions, *IEEE Trans. Automat. Contrl.*, Vol. **AC-20**, No. 5, pp. 596–603, 1975.
8. Chen, C. F., Tsay, Y. T. and Wu, T. T., Walsh operational matrices for fractional calculus and their application to distributed systems, *J. Franklin Inst.*, Vol. **303**, No. 3, pp. 267–284, 1977.
9. Ogata, K., *Modern Control Engineering* (5th Ed.), Prentice Hall, New Delhi, 2010.
10. Harmuth, H. F., *Transmission of Information by Orthogonal Functions* (2nd Ed.), Springer-Verlag, Berlin, 1972.
11. Guillemin, E. A., *The Mathematics of Circuit Analysis*, Oxford, Calcutta, 1967.
12. Walsh, J. L., A closed set of normal orthogonal functions, *Am. J. Math.*, Vol. **45**, pp. 5–24, 1923.

13. Moulden, T. H. and Scott, M. A., Walsh spectral analysis for ordinary differential equations: Part I—Initial value problems, *IEEE Trans. Circuits Syst.*, Vol. **35**, No. 6, pp. 742–745, 1988.
14. Deb, A. and Fountain, D. W., A note on oscillations in Walsh domain analysis of first order systems, *IEEE Trans Circuits Syst.*, Vol. **38**, No. 8, pp. 945–948, 1991.
15. Jiang, J. H. and Schaufelberger, W., *Block Pulse Functions and their Application in Control System*, LNCIS 179, Springer-Verlag, Berlin, 1992.

4

Analysis of Pulse-Fed Single-Input Single-Output Systems

The pulse-width modulation (PWM) technique using thyristors and gate turn-off thyristor (GTO) switches is a popular method for controlling output DC voltage/current from a fixed supply in a lossless manner. There are many schemes for attaining such control, encompassing numerous application areas such as traction, uninterruptible power supply (UPS), and voltage regulators [1,2]. The complexities of operation of such DC-to-DC converters are compounded by the nature of load that may be nonlinear in some cases. To have a complete knowledge of the circuits and to predict the performances of the devices, various analytical techniques are proposed with the basic objective to know the total history of variation of all significant variables with respect to time.

Further, it will also enable us to determine the ratings of different semiconductor components involved. Hence, it is necessary to know the average and root mean square quantities of the physical variables such as current. A number of efforts by different techniques have been expanded on the analysis of DC-to-DC converters operating on PWM techniques feeding a motor or a higher order load with many time constants [3]. The most common of these methods is the Laplace transform technique. A straightforward application of Laplace transform to higher order systems has limitations with respect to the inconvenience caused in obtaining the inverse transform of the output [4]. Further, for different kinds of inputs to a single system, the inverse transforms have to be computed each time separately, as already mentioned in Chapter 3.

In case of nonlinear loads, the Laplace transform approach assumes a linearized model to obtain the solution considering perturbation around a stable operating point. Fourier series expansion and state variable technique are also applied to such loads, assuming them to be piecewise linear in nature [5,6].

The generality of state variable method of system representation also found its usefulness in the analysis of power electronic (PE) circuits. However, the different modes of operation of switching devices pose constraints on the selection of state vectors [7,8]. The repetitive nature of operation of the switching devices in a PE circuit has attracted the application of the switching function technique, in which very complex waveforms can be simulated by switching functions for the purpose of analysis [9–11]. However, no general procedure using this method has yet been proposed.

With the recent widespread applications of microprocessors for controlling digitally the PE circuits, analytical techniques by the sampled data method [12] and modified z-transform method [13] have gained momentum. But these methods have their own limitations while handling systems having a large number of poles and zeros.

While these methods yield satisfactory results, their drawback lies in the fact that for a particular system subjected to different kinds of input waveforms, for example, a pulse-width modulated wave or a phase-controlled wave, the analysis has to be repeated all over for each waveform. Also, these methods are not very convenient for computer manipulation.

A new method of analysis of pulse-width modulated systems is presented in this section, which is carried out in "sequency domain" instead of frequency domain as defined by Harmuth [14]. The principle of the proposed method was outlined in Chapter 3.

In this chapter, we study the response of an ideal chopper-fed system [15–20] by Walsh domain analysis. To achieve this goal, we start with the study of a first-order and a second-order system where the driving function is a single pulse, a pulse pair, or an alternating pulse pair. The method is extended to the analysis and simulation of a PWM chopper feeding an RL load. Further, an application of the proposed method is established for a chopper-fed DC series motor load.

4.1 Analysis of a First-Order System

4.1.1 Single-Pulse Input

The first-order system discussed in Section 3.4 may now be investigated with single-pulse input. A square wave of certain amplitude, width, and frequency is defined here as a pulse train. A single-pulse input is therefore just a single square pulse having a certain amplitude and width. Therefore, an input function of the following nature is chosen:

$$r(t) = u(t) - u(t - t_1)$$

This is nothing but a pulse block of unit height and t_1-second duration starting from origin shown in Figure 4.1a. Considering a total interval of 4 seconds, let us assume $t_1 = 0.5$ seconds so that the input pulse covers one-eighth of the total interval. Referring to Figure 2.8, we can see that the pulse input chosen is actually the first member of the block pulse function family, the only difference being the scaling factor $\lambda = 4$. Hence, the input is

$$r(t) = u(t) - u(t - 0.5) \tag{4.1}$$

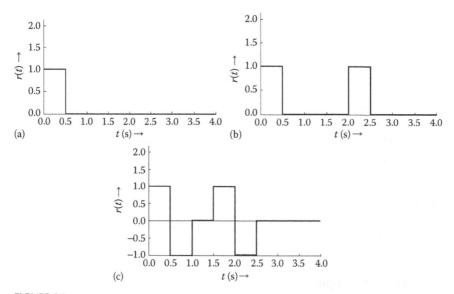

FIGURE 4.1
(a) Single-pulse input, (b) pulse-pair input and (c) alternating double-pulse input for $m = 8$ and $T = 4$ seconds.

In Walsh domain, considering $m = 8$, $r(t)$ is represented as

$$r(t) \approx \frac{1}{8}[1 \quad 1 \quad 1 \quad 1 \quad 1 \quad 1 \quad 1 \quad 1]\Phi_{(8)} = \mathbf{R}\,\Phi_{(8)} \qquad (4.2)$$

Since we consider the same system as in Section 3.4, we use the same operational transfer function **WOTF1** to determine the output response.
We make use of Equation 3.41 to write

$$\mathbf{C}\Phi_{(8)} = \mathbf{WOTF1}\ \mathbf{R}\,\Phi_{(8)}$$

or $\qquad\qquad\qquad\qquad\qquad\qquad\qquad\qquad\qquad\qquad\qquad\qquad$ (4.3)

$$\mathbf{C}\Phi_{(8)} = \frac{1}{8}[1 \quad 1 \quad 1 \quad 1 \quad 1 \quad 1 \quad 1 \quad 1]\,\mathbf{WOTF1}\ \Phi_{(8)}$$

Substituting **WOTF1** from Equation 3.47, the output response in Walsh domain is

$$\mathbf{C}\Phi_{(8)} = [0.1222 \quad 0.0846 \quad 0.0354 \quad 0.0178 \quad -0.0007 \quad -0.0101 \quad -0.0224 \quad -0.0268]\,\Phi_{(8)} \quad (4.4)$$

The Walsh response is shown in Figure 4.2 along with the actual time response (Program B.13 in Appendix B).

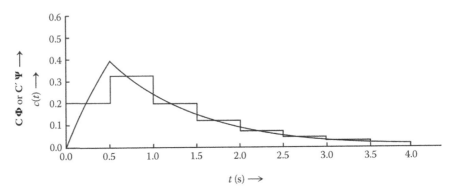

FIGURE 4.2
Time response of a first-order system with single-pulse input and its equivalent Walsh/block pulse representation for $m = 8$ and $T = 4$ seconds.

4.1.2 Pulse-Pair Input

Now we consider a pulse pair fed to the same system defined by Equation 3.45. This pulse-pair input is shown in Figure 4.1b. In our specified domain, namely, 4 seconds, let us assume that we have two pulse blocks having pulse width of 0.5 seconds each and a pulse frequency or repetition rate of 2 seconds.

This input may be expressed in Walsh domain as

$$\mathbf{R}\Phi_{(8)} = \frac{1}{4}[1 \quad 0 \quad 1 \quad 0 \quad 1 \quad 0 \quad 1 \quad 0]\Phi_{(8)} \tag{4.5}$$

Hence, the output response in the Walsh domain is given by

$$\mathbf{C}\Phi_{(8)} = \frac{1}{4}[1 \quad 0 \quad 1 \quad 0 \quad 1 \quad 0 \quad 1 \quad 0]\mathbf{WOTF1}\,\Phi_{(8)} \tag{4.6}$$

In Walsh domain, the output response is

$$\mathbf{C}\Phi_{(8)} = [0.2256 \quad -0.0188 \quad 0.0620 \quad -0.0088 \quad -0.0061 \quad -0.0047 \quad -0.0470 \quad -0.0022]\Phi_{(8)} \tag{4.7}$$

The actual response and the Walsh response (Program B.14 in Appendix B) are shown in Figure 4.3.

4.1.3 Alternating Double-Pulse Input

Consider the system input slightly more complex as shown in Figure 4.1c. Here the total time considered is $T = 4$ seconds. In Walsh domain, the input is represented as

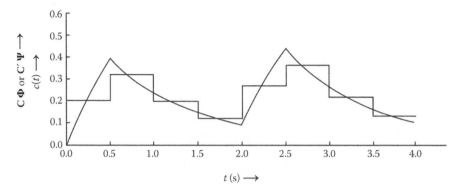

FIGURE 4.3
Time response of a first-order system with pulse-pair input and its equivalent Walsh/block pulse representation for $m = 8$ and $T = 4$ seconds.

$$r(t) = u(t) - 2u(t-1) + u(t-2) + u(t-3) - 2u(t-4) + u(t-5)$$

$$\approx \left[0 \quad \frac{1}{4} \quad -\frac{1}{4} \quad 0 \quad 0 \quad \frac{1}{4} \quad \frac{1}{4} \quad \frac{1}{2} \right] \Phi_{(8)} = \mathbf{R}\,\Phi_{(8)} \tag{4.8}$$

with eight basis functions and a scaling constant $\lambda = 4$.

Here, for the same first-order system, we use the same **WOTF1** from Equation 3.47.

Thus,

$$\mathbf{C}\Phi_{(8)} = \left[0 \quad \frac{1}{4} \quad -\frac{1}{4} \quad 0 \quad 0 \quad \frac{1}{4} \quad \frac{1}{4} \quad \frac{1}{2} \right] \mathbf{WOTF1}\,\Phi_{(8)} \tag{4.9}$$

Substituting **WOTF1** from Equation 3.47, the output response in Walsh domain is

$$\mathbf{C}\Phi_{(8)} = \left[0.0105 \quad 0.0683 \quad 0.0491 \quad 0.0321 \quad 0.0026 \quad -0.0454 \quad 0.0748 \quad 0.0080 \right] \Phi_{(8)} \tag{4.10}$$

Figure 4.4 (Program B.15 in Appendix B) shows the time response of the first-order system with alternating double-pulse input and its equivalent Walsh domain representation.

In Figure 4.4, the deviation between the two curves shows that for oscillatory output functions, to increase accuracy, we need to use more number of component basis functions. In the above example, if we analyze the system with $m = 16$, the result will trace the exact solution more closely.

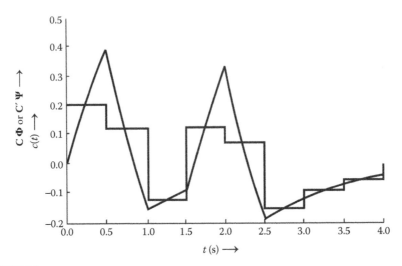

FIGURE 4.4
Time response of a first-order system with alternating double-pulse input and its equivalent Walsh domain representation for $m = 8$ and $T = 4$ seconds.

4.2 Analysis of a Second-Order System

4.2.1 Single-Pulse Input

We consider a second-order system having a transfer function given by Equation 3.52. The driving function is a single pulse of 1-second duration in an interval of 8 seconds. Thus, the pulse covers one-eighth of the total interval as in the case of a first-order system discussed in Section 4.1.1.

Hence, the input in Figure 4.1a is simply doubled in time scale and represented by

$$r(t) = u(t) - u(t-1) \approx \frac{1}{8}[1 \quad 1 \quad 1 \quad 1 \quad 1 \quad 1 \quad 1 \quad 1]\Phi_{(8)} = \mathbf{R}\,\Phi_{(8)} \qquad (4.11)$$

with eight basis Walsh functions and a scaling constant $\lambda = 8$.

Using **WOTF2** from Equation 3.54, we obtain the output response in Walsh domain as

$$\mathbf{C}\Phi_{(8)} = [0.1062\ 0.2390\ -0.0352\ 0.0099\ -0.0041\ 0.0023\ -0.0432\ -0.1137]\Phi_{(8)} \qquad (4.12)$$

The Walsh domain solution along with the actual solution is shown in Figure 4.5 (Program B.16 in Appendix B).

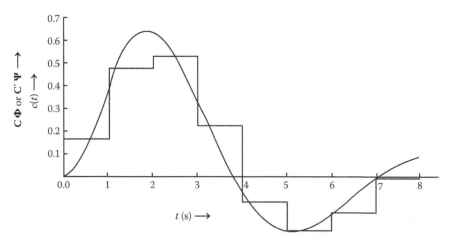

FIGURE 4.5
Time response of a second-order system with single-pulse input and its equivalent Walsh/block pulse representation for $m = 8$ and $T = 8$ seconds.

4.2.2 Pulse-Pair Input

Now the same second-order system defined by Equation 3.52 is excited with a pulse-pair input. Again, we consider an interval of 8 seconds and each pulse of the input pulse-pair is of 1-second duration with a pulse repetition rate of 4 seconds.

Proceeding in a manner as before, we make use of **WOTF2** to obtain the output response $\mathbf{C}\Phi_{(8)}$ in Walsh domain as

$$\mathbf{C}\Phi_{(8)} = [0.2788 \quad 0.0664 \quad -0.0478 \quad 0.0226 \quad -0.0049 \quad 0.0032 \quad -0.1217 \quad -0.0352]\Phi_{(8)} \quad (4.13)$$

The Walsh domain response along with the actual solution is shown in Figure 4.6 (Program B.17 in Appendix B).

4.2.3 Alternating Double-Pulse Input

In this case, the total time considered is $T = 8$ seconds. The system input shown in Figure 4.1c is doubled in time scale and represented in Walsh domain as

$$r(t) \approx \left[0 \quad \frac{1}{4} \quad -\frac{1}{4} \quad 0 \quad 0 \quad \frac{1}{4} \quad \frac{1}{4} \quad \frac{1}{2} \right]\Phi_{(8)} = \mathbf{R}\,\Phi_{(8)} \quad (4.14)$$

with $m = 8$ and a scaling constant $\lambda = 8$.

Using **WOTF2**, we determine the output response in Walsh domain as

$$\mathbf{C}\Phi_{(8)} = [-0.0156 \quad 0.1097 \quad 0.1120 \quad 0.0332 \quad 0.0065 \quad 0.0035 \quad -0.0282 \quad -0.0599]\Phi_{(8)} \quad (4.15)$$

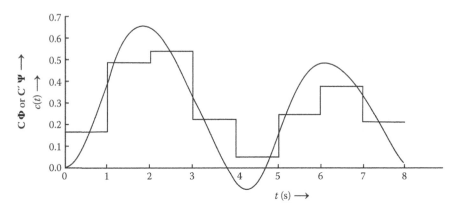

FIGURE 4.6
Time response of a second-order system with pulse-pair input and its equivalent Walsh/block pulse representation for $m = 8$ and $T = 8$ seconds.

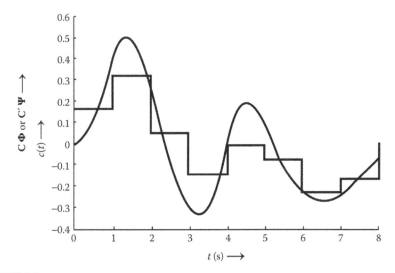

FIGURE 4.7
Time response of a second-order system with alternating double-pulse input and its equivalent Walsh representation for $m = 8$ and $T = 8$ seconds.

The Walsh domain response along with the actual solution is shown in Figure 4.7 (Program B.18 in Appendix B).

Here also, we need to use more number of component basis functions to increase the accuracy of the oscillatory output function. If we analyze the system with $m = 16$ or 32, the result will be more close to the exact solution.

4.3 Pulse-Width Modulated Chopper System

In what follows, a chopper circuit, producing ideal square waves, has been investigated. An ideal chopper acts as a lossless switch, supplying voltage to the load during "on" condition of the switch. For an inductive load, the load current is generally allowed to freewheel through a diode connected across the load as shown in Figure 4.8a. The switch of the chopper may be a thyristor along with its commutation circuit or a GTO.

For designing an ideal thyristor chopper, the commutation circuit is not allowed to charge or discharge through load. To control the average load current, voltage PWM technique is generally adopted where the switch is allowed to operate at a fixed repetition rate (analogous to frequency) and variable duty circle (i.e., "on" time). For a chopper producing an ideal square wave, a commutation circuit usually consists of two auxiliary thyristors and associated L–C charging circuits [21]. Triggers to main and auxiliary thyristors are produced by a voltage-to-pulse width converter, the output of which is processed to generate three trigger signals. The basic scheme is shown in Figure 4.8b.

4.3.1 Case I: Stepwise PWM

4.3.1.1 Walsh Function Representation of Significant Current Variables

In a chopper circuit, the values of average and root mean square (rms) load currents along with the average and rms currents in the main switching

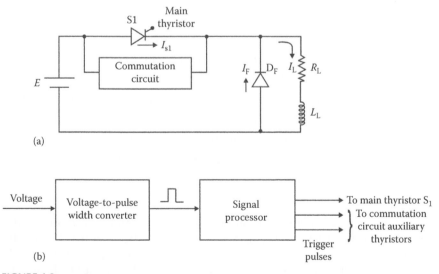

(a)

(b)

FIGURE 4.8
(a) Chopper system with freewheeling diode and (b) basic triggering scheme. (Reprinted from Deb, A. and Datta, A. K., *Int. J. Electron.*, 62, 449–459, 1987. With permission.)

element and through the freewheeling diode are of much interest since these dictate the selection of semiconductor components. This has been investigated for an inductive load having $R_L = 31\ \Omega$ and $L_L = 0.17$ H.

The transfer function of the system, considering the normalized load current I_L as the output variable, is given by

$$G(s) = \frac{31}{31 + 0.17s}$$

Assuming that the output values are to be calculated in a time interval of 0.08 seconds, the Walsh operational transfer function of the system (say **WOTF3**) in the scaled domain may be written as

$$\mathbf{WOTF3} = 31\ \mathbf{I}\left[31\ \mathbf{I} + \left(\frac{0.17}{0.08}\right)\mathbf{D}\right]^{-1} \qquad (4.16)$$

where:
  I is a unit matrix of order 8, since we choose $m = 8$

By substituting the operational matrix for differentiation $\mathbf{D}_{(8)}$ from Equation 3.19, **WOTF3** could be evaluated as

WOTF3 =

$$14.588 \begin{bmatrix} 0.06385 & -0.00470 & -0.00468 & -0.00468 & -0.00428 & -0.00428 & -0.00427 & -0.00427 \\ 0.00470 & 0.05445 & 0.00468 & -0.01404 & 0.00428 & -0.01285 & 0.00427 & -0.01280 \\ 0.00468 & 0.00468 & 0.03578 & 0.00466 & 0.00427 & 0.00427 & -0.02988 & 0.00425 \\ -0.00468 & 0.01404 & -0.00466 & 0.04509 & -0.00427 & 0.01280 & -0.00425 & -0.02138 \\ 0.00428 & 0.00428 & 0.00427 & 0.00427 & 0.00390 & 0.00390 & 0.00389 & 0.00389 \\ -0.00428 & 0.01285 & -0.00427 & 0.01280 & -0.00390 & 0.01172 & -0.00389 & 0.01167 \\ -0.00427 & -0.00427 & 0.02988 & -0.00425 & -0.00389 & -0.00389 & 0.02724 & -0.00387 \\ 0.00427 & -0.01280 & 0.00425 & 0.02138 & 0.00389 & -0.01167 & 0.00387 & 0.01950 \end{bmatrix} \qquad (4.17)$$

In the interval under consideration, we choose three different input simulations **A**, **B**, and **C** shown in Figure 4.9. It is apparent from Figure 4.9 that the pulse width of the input waveform **A** is modulated by steps of 0.01 seconds (=25% of the repetition rate T_R, 0.04 seconds) to obtain inputs **B** and **C**. The very nature of the input waves enables us to represent them in terms of Walsh functions as outlined in Section 3.3. Thus,

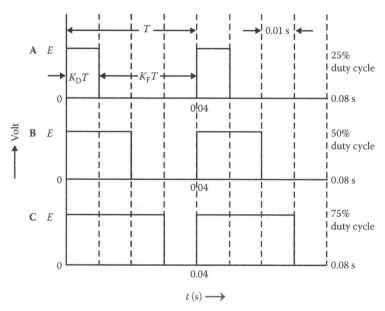

FIGURE 4.9

Three inputs **A**, **B**, and **C** for stepwise PWM ($T = 0.08$ seconds). (Reprinted from Deb, A. and Datta, A. K., *Int. J. Electron.*, 62, 449–459, 1987. With permission.)

$$
\left.
\begin{aligned}
&\text{Input } \mathbf{A}\,(\text{pulse width } 25\%) \\[4pt]
&\qquad \approx \mathbf{R1}\,\Phi_{(8)}(t) \\[4pt]
&\qquad = \frac{1}{4}\,[1 \quad 0 \quad 1 \quad 0 \quad 1 \quad 0 \quad 1 \quad 0]\Phi_{(8)}(t) \\[6pt]
&\text{Input } \mathbf{B}\,(\text{pulse width } 50\%) \\[4pt]
&\qquad \approx \mathbf{R2}\,\Phi_{(8)}(t) \\[4pt]
&\qquad = \frac{1}{2}\,[1 \quad 0 \quad 1 \quad 0 \quad 0 \quad 0 \quad 0 \quad 0]\Phi_{(8)}(t) \\[6pt]
&\text{Input } \mathbf{C}\,(\text{pulse width } 75\%) \\[4pt]
&\qquad \approx \mathbf{R3}\,\Phi_{(8)}(t) \\[4pt]
&\qquad = \frac{1}{4}\,[3 \quad 0 \quad 1 \quad 0 \quad 1 \quad 0 \quad -1 \quad 0]\Phi_{(8)}(t)
\end{aligned}
\right\}
\tag{4.18}
$$

Now Equations 4.16 through 4.18 could be used to evaluate $\mathbf{C}\,\Phi_{(8)}(t)$ following Equation 3.41. Thus, the output load current waveforms $\mathbf{C1}\,\Phi(t)$, $\mathbf{C2}\,\Phi(t)$, and $\mathbf{C3}\,\Phi(t)$, corresponding to the pulse inputs $\mathbf{R1}\,\Phi(t)$, $\mathbf{R2}\,\Phi(t)$, and $\mathbf{R3}\,\Phi(t)$, respectively, are given by

$$\left.\begin{array}{l} \textbf{C1 } \Phi_{(8)}(t) = \frac{1}{4}\,[1 \quad 0 \quad 1 \quad 0 \quad 1 \quad 0 \quad 1 \quad 0]\textbf{WOTF3 } \Phi_{(8)}(t) \\[3mm] \textbf{C2 } \Phi_{(8)}(t) = \frac{1}{2}\,[1 \quad 0 \quad 1 \quad 0 \quad 0 \quad 0 \quad 0 \quad 0]\textbf{WOTF3 } \Phi_{(8)}(t) \\[3mm] \textbf{C3 } \Phi_{(8)}(t) = \frac{1}{4}\,[3 \quad 0 \quad 1 \quad 0 \quad 1 \quad 0 \quad -1 \quad 0]\textbf{WOTF3 } \Phi_{(8)}(t) \end{array}\right\} \quad (4.19)$$

From the above output equations, it is apparent that the resultant coefficient matrix of $\Phi_{(8)}$ will always be a (1×8) row matrix whose elements are the coefficients of the corresponding eight Walsh functions. Superposition of these eight-component Walsh functions in each case will generate the stepped waveform of the output $\textbf{C}\Phi_{(8)}(t)$.

To attain a better insight into the results, we convert the Walsh function components to block pulse functions with the help of Equation 2.16, and then get $\textbf{C1}'\ \Psi(t)$, $\textbf{C2}'\ \Psi(t)$, and $\textbf{C3}'\ \Psi(t)$ as

$$\left.\begin{array}{l} \textbf{C1}'\ \Psi_{(8)}(t) = \frac{1}{4}\,[1 \quad 0 \quad 1 \quad 0 \quad 1 \quad 0 \quad 1 \quad 0]\{\textbf{WOTF3}\}\ \textbf{W}\ \Psi_{(8)}(t) \\[3mm] \textbf{C2}'\ \Psi_{(8)}(t) = \frac{1}{2}\,[1 \quad 0 \quad 1 \quad 0 \quad 0 \quad 0 \quad 0 \quad 0]\{\textbf{WOTF3}\}\ \textbf{W}\ \Psi_{(8)}(t) \\[3mm] \textbf{C3}'\ \Psi_{(8)}(t) = \frac{1}{4}\,[3 \quad 0 \quad 1 \quad 0 \quad 1 \quad 0 \quad -1 \quad 0]\{\textbf{WOTF3}\}\ \textbf{W}\ \Psi_{(8)}(t) \end{array}\right\} \quad (4.20)$$

Hence, the general expression for the normalized load current in terms of block pulse functions is

$$\textbf{C}'\ \Psi_{(8)}(t) = [c_0' \quad c_1' \quad c_2' \quad c_3' \quad c_4' \quad c_5' \quad c_6' \quad c_7']\ \Psi_{(8)}(t) \quad (4.21)$$

which is the time response of the normalized load current I_L in the interval 0 to 0.08 seconds, and c_0', c_1',... are the piecewise constant amplitudes in each subinterval of 0.01 second.

However, substitution of **WOTF3** from Equation 4.17 and **W** from Equation 3.42 in Equation 4.20 yields the output response that is shown in Figure 4.10 (Program B.19 in Appendix B) along with the exact waveforms.

4.3.1.2 Determination of Normalized Average and rms Currents through Load and Semiconductor Components

To determine the normalized average and rms currents through load and semiconductor components, we use Equation 4.21. In this equation, it is apparent that the eight coefficients, namely, c_0', c_1', c_2',..., c_7' give the normalized

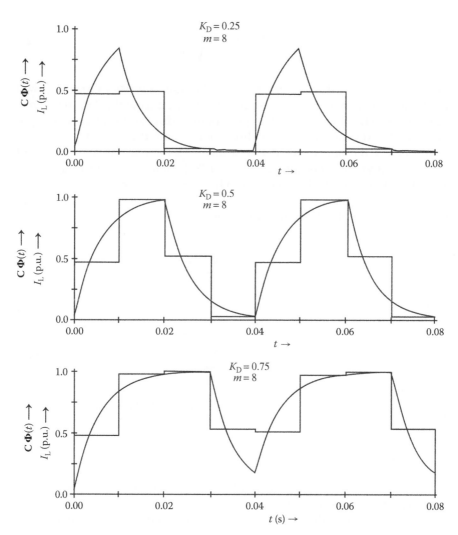

FIGURE 4.10
Exact and Walsh/block pulse waveforms of the normalized load current for $m = 8$, $T = 0.08$ seconds, and $K_D = 0.25$, 0.50, and 0.75.

average load current in each consecutive subinterval of 0.01 second starting from $t = 0$. Thus, the coefficients c_0', c_1', c_2',... directly give the average amplitudes of the output waveform in different subintervals.

To obtain the normalized load current averaged over the entire period of 0.08 seconds ($=T$), it follows from Equation 4.21 that

$$I_{L(av)} = \mathbf{C}' \mathbf{\Psi}_{(8)}(t)_{(av)} = \frac{1}{8} \sum_{n=0}^{7} c_n' \tag{4.22}$$

and the corresponding rms value is

$$I_{L(rms)} = \mathbf{C}' \, \Psi_{(8)}(t)_{(rms)} = \sqrt{\left[\frac{1}{8} \sum_{n=0}^{7} c_n'^2 \right]} \qquad (4.23)$$

Referring to Figure 4.9, it is obvious that the main thyristor current I_{s1} will exist only when an input pulse is present, that is, during the "on" time only. But, the freewheeling current I_F is present only during the absence of input pulse, that is, during the "off" condition of the main thyristor s1.

In view of the above, we define a logic operator δ that will represent the "main thyristor condition": "on" and "off" states of main thyristor s1. Therefore, when $\delta = 1$, s1 is "on," and when $\delta = 0$, s1 is "off."

Hence, the average and rms values of the normalized device current over the whole period of 0.08 seconds are given by

$$I_{s1(av)} = \frac{1}{8} \sum_{n=0}^{7} c_n' \delta \qquad (4.24)$$

and

$$I_{s1(rms)} = \sqrt{\left[\frac{1}{8} \sum_{n=0}^{7} c_n'^2 \delta^2 \right]} \qquad (4.25)$$

For 25% duty cycle, input **A** in Figure 4.9, Equation 4.24 becomes

$$I_{s1(0.25)\,(av)} = \frac{1}{8} [c_0' + c_4'] \qquad (4.26)$$

since $\delta = 1$ only for $n = 0$ and 4, and for the other six coefficients, $n = 1, 2, 3, 5, 6,$ and 7, the logic operator $\delta = 0$.

For the same input, the rms value of the normalized main thyristor current, using Equation 4.25, is given by

$$I_{s1(0.25)\,(rms)} = \sqrt{\left[\frac{1}{8} [c_0'^2 + c_4'^2] \right]} \qquad (4.27)$$

For freewheeling current, we have expressions identical to Equations 4.24 and 4.25; only the logic operator δ in this case will be complementary of δ or $\bar{\delta}$. As the freewheeling current can only flow during the "off" condition of the main thyristor s1, considering $\bar{\delta}$.

$$I_{F(0.25)(av)} = \frac{1}{8} [c_1' + c_2' + c_3' + c_5' + c_6' + c_7'] \qquad (4.28)$$

TABLE 4.1

Normalized Average Currents ($R_L = 31\ \Omega$, $L_L = 0.17$ H, $m = 8$, $T = 0.08$ seconds)

Duty Cycle K_D (%)	Normalized Load Current $I_{L(av)}$		Normalized Device Current $I_{s1(av)}$		Normalized Freewheeling Current $I_{F(av)}$	
	Exact	Walsh	Exact	Walsh	Exact	Walsh
25	0.2497	0.2500	0.1352	0.1192	0.1145	0.1307
50	0.4982	0.4999	0.3682	0.3634	0.1301	0.1365
75	0.7390	0.7470	0.6244	0.6161	0.1145	0.1307

TABLE 4.2

Normalized rms Currents ($R_L = 31\ \Omega$, $L_L = 0.17$ H, $m = 8$, $T = 0.08$ seconds)

Duty Cycle K_D (%)	Normalized Load Current $I_{L(rms)}$		Normalized Device Current $I_{s1(rms)}$		Normalized Freewheeling Current $I_{F(rms)}$	
	Exact	Walsh	Exact	Walsh	Exact	Walsh
25	0.3677	0.3453	0.2950	0.2385	0.2196	0.2498
50	0.6067	0.6027	0.5506	0.5432	0.2550	0.2613
75	0.7897	0.7849	0.7465	0.7401	0.2574	0.2615

and

$$I_{F(0.25)\ (rms)} = \sqrt{\frac{1}{8}\left[c_1'^2 + c_2'^2 + c_3'^2 + c_5'^2 + c_6'^2 + c_7'^2 \right]} \tag{4.29}$$

Similarly, for other duty cycles (i.e., 50% and 75% pulse inputs), different normalized average and rms currents could be computed.

Using equations given in the following section, the exact values of the normalized average and rms currents for the above inputs are calculated and the results are summarized in Tables 4.1 and 4.2 (Program B.19 in Appendix B). Table 4.1 compares the normalized average currents obtained by exact computation and Walsh operational method, whereas Table 4.2 compares similar results for the normalized rms current. The results obtained by the above two methods are fairly close as evident from the tables.

4.3.1.3 Determination of Exact Normalized Average and rms Current Equations Considering Switching Transients

For an ideal DC chopper, the voltage and current waveforms in an RL load, considering switching transients, are shown in Figure 4.11. We assume that the voltage across the load is square, and the commutating interval is negligible in comparison with the time period T_R.

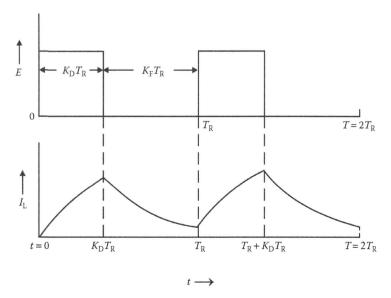

FIGURE 4.11
Voltage and current waveforms of an ideal chopper. (Reprinted from Deb, A. and Datta, A. K., *Int. J. Electron.*, 62, 449–459, 1987. With permission.)

Then, the expressions for the normalized main thyristor current (I_{s1}) and freewheeling current (I_F) are given by

$$\frac{I_{s1}R_L}{E} = I_{s1} = \left[1 - \exp\left(-\frac{R_L t}{L_L}\right)\right] u(t)$$

$$+ \left[\{1 - \exp(-\alpha K_D)\} \, \exp(-\alpha K_F) \, \exp\left\{-\frac{R_L(t - T_R)}{L_L}\right\} \right] \qquad (4.30)$$

$$+ \left[1 - \exp\left\{-\frac{R_L(t - T_R)}{L_L}\right\} \right] u(t - T_R)$$

and

$$\frac{I_f R_L}{E} = I_F = \left[\{1 - \exp(-\alpha K_D)\} \, \exp\left\{-\frac{R_L(t - K_D T_R)}{L_L}\right\} \right] u(t - K_D T_R)$$

$$+ \left[\{1 - \exp(-\alpha K_D)\} \, \{\exp(-\alpha) + 1\} \, \exp\left\{\frac{-R_L(t - T_R - K_D T_R)}{L_L}\right\} \right] \qquad (4.31)$$

$$u(t - T_R - K_D T_R)$$

where:

$$\alpha = \left(\frac{T_R R_L}{L_L}\right) = \frac{T_R}{T_L}$$

$T_L = L_L/R_L$ is the load time constant
$T_R = 0.04$ second is the repetition rate of the input waveform
K_D is the duty cycle of chopper
K_F is the freewheeling cycle of chopper $= (1 - K_D)$

From Equation 4.24, the normalized average device current is given by

$$I_{s1(av)} = \frac{1}{2T_R}\left[\int_0^{K_D T_R}\left\{1-\exp\left(-\frac{R_L t}{L_L}\right)\right\}dt + \int_{T_R}^{T_R+K_D T_R}\left[\{1-\exp(-\alpha K_D)\}\right.\right.$$

$$\exp(-\alpha K_F)\exp\left\{-R_L\frac{(t-T_R)}{L_L}\right\} + \qquad (4.32)$$

$$\left.\left.\left[1-\exp\left\{-R_L\frac{(t-T_R)}{L_L}\right\}\right]\right]dt\right]$$

After integration and simplification, we have

$$I_{s1(av)} = K_D - \left[\frac{\{1-\exp(-\alpha K_D)\}}{\alpha}\right] + [1-\exp(-\alpha K_D)]^2\frac{\exp(-\alpha K_F)}{2\alpha} \qquad (4.33)$$

Similarly, from Equation 4.31, the normalized average freewheeling current is given by

$$I_{F(av)} = \frac{1}{2T_R}\left[\int_{K_D T_R}^{T_R}[1-\exp(-\alpha K_D)]\exp\left\{-R_L\frac{(t-K_D T_R)}{L_L}\right\}dt + \right.$$

$$\int_{T_R+K_D T_R}^{2T_R}\left[\{1-\exp(-\alpha K_D)\}\exp(-\alpha)+\{1-\exp(-\alpha K_D)\}\right] \qquad (4.34)$$

$$\left.\exp\left\{-R_L\frac{(t-T_R-K_D T_R)}{L_L}\right\}dt\right]$$

After integration and simplification, we get

$$I_{F(av)} = \{1-\exp(-\alpha K_D)\}\{1-\exp(-\alpha K_F)\}\left[\frac{\{2+\exp(-\alpha)\}}{2\alpha}\right] \qquad (4.35)$$

Hence, the normalized average load current is given by

$$I_{L(av)} = I_{s1(av)} + I_{F(av)} \tag{4.36}$$

So, using Equations 4.33, 4.35, and 4.36, we can compute $I_{s1(av)}$, $I_{F(av)}$, and $I_{L(av)}$. From Equation 4.30, the normalized rms thyristor current or device is

$$I_{s1\,(rms)} = \sqrt{\begin{bmatrix} \dfrac{1}{2T_R} \begin{bmatrix} \displaystyle\int_0^{K_D T_R} \left[1 - \exp\left(-\dfrac{R_L t}{L_L}\right)\right]^2 dt + \\[2em] \displaystyle\int_{T_R}^{T_R + K_D T_R} \left[\{1 - \exp(-\alpha K_D)\} \exp(-\alpha K_F) \exp\left\{-\dfrac{R_L(t - T_R)}{L_L}\right\} + \right. \\[2em] \left[1 - \exp\left\{-R_L \dfrac{(t - T_R)}{L_L}\right\} \right]^2 dt \end{bmatrix} \end{bmatrix}} \tag{4.37}$$

After integration and simplication, we have

$$I_{s1\,(rms)} = \sqrt{\left[K_D - 2A + 2B - 2AB\alpha \, \exp(-\alpha K_F) + \alpha^2 A^2 B \, \exp(-2\alpha K_F) + \alpha A^2 \, \exp(-\alpha K_F) \right]} \tag{4.38}$$

where:

$$A = \frac{[1 - \exp(-\alpha K_D)]}{\alpha}$$

$$B = \frac{[1 - \exp(-2\alpha K_D)]}{4\alpha}$$

In a similar manner, the normalized rms freewheeling current may be derived from Equation 4.31 as

$$I_{F\,(rms)} = \frac{1}{2}\sqrt{\left[\alpha A^2 \{1 - \exp(-2\alpha K_F)\} \{2 + 2\exp(-\alpha) + \exp(-2\alpha)\} \right]} \tag{4.39}$$

From Equations 4.38 and 4.39, the normalized rms load current is obtained as

$$I_{L(rms)} = \sqrt{\left[I_{s1(rms)}^2 + I_{F(rms)}^2 \right]} \tag{4.40}$$

Equations 4.33, 4.35, 4.36, and 4.38 through 4.40 give the exact solutions for different normalized currents considering switching transients in the interval

$T = 2T_R$. For the circuit under investigation, $T_R = 0.04$ seconds, $T = 0.08$ seconds, $T_L = 0.005484$ seconds, $\alpha = 7.2941$, and the different values of K_D from 0.1 to 1.0 were chosen. Using these values, the exact solutions for various normalized average and rms currents are obtained from Equations 4.33, 4.35, 4.36, and 4.38 through 4.40, and these are used to find out the exact solutions tabulated in Tables 4.1 and 4.2.

4.3.2 Case II: Continuous PWM

This section introduces continuous PWM in contrast to the stepwise modulation presented in Section 4.3.1. A continuously pulse-width modulated PE system [21] is investigated and its output response is determined.

A chopper-fed RL load is considered as an illustration of the presented method in which the average and rms values of important circuit variables are obtained and compared with the exact values.

In this analysis, we use 16 basis Walsh functions instead of 8 that were used in all the previous analyses. This evidently improves the accuracy of the computed results that are found to be in good agreement with the exact solutions.

The PWM is effected by shifting the trailing edge of an input square pulse while keeping its leading edge fixed, with a constant repetition rate. In Figure 4.12a, we consider a pulse input in a time interval of T seconds, which is divided into m equal subintervals of T/m seconds each, where m is of the form $m = 2^p$, p being a positive integer.

Let the repetition rate of the input square wave be $T/2$ seconds. Evidently, only two pulses are considered initially, each of which could cover any portion of the repetition period $T/2$. However, each pulse may consist of n number of subintervals $[n < (m/2)]$ of T/m-second duration and a fraction x $(0 < x < 1)$ of the $(n + 1)$th subinterval. Thus, xT/m seconds of the $(n + 1)$th subinterval is covered by the input pulse, while for the rest $[(1 - x)T/m]$ seconds, the input is 0. Therefore, the total time span of a single input pulse is $[(n + x)T/m]$ seconds.

If the amplitude of the input waveform is A, the area covered by each pulse slice in the $(n + 1)$th subinterval is $(AxT)/m$ as shown by the shaded portion in Figure 4.12a. At this point, we fall back on the area-preserving technique to transform the waveform of Figure 4.12a to that of Figure 4.12b, which is suitable for representation by Walsh functions. Since Walsh functions are piecewise constant and a complete set guided by Parseval's theorem, such area-preserving transformation is safe in the case as for most of similar cases. Hence, keeping the shaded area of Figure 4.12a constant, we convert the pulse slice into another pulse slice of different shape but of equal area so as to have a width of T/m seconds and a height Ax. Thus, the area of the newly shaped pulse slice is unchanged and equal to $(AxT)/m$. This is illustrated in Figure 4.12b, where the reshaped pulse slice is shown by the shaded area. The modified input waveform of Figure 4.12b can now be represented

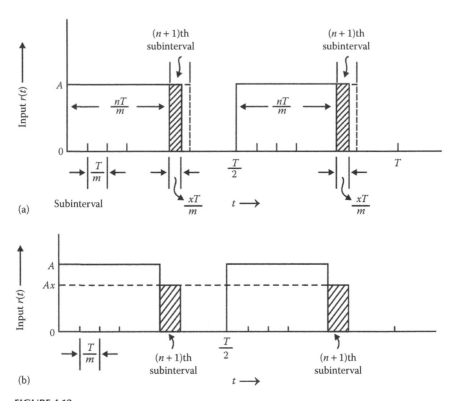

FIGURE 4.12
(a) A typical pulse-train input where each pulse covers n subinterval and a fraction x of the $(n + 1)$th subinterval and (b) the same pulse-train input after area-preserving transformation. (Reprinted from Deb, A. and Datta, A. K., *Int. J. Sys. Sci.*, 23, 151–166, 1992. With permission.)

by Walsh functions. To do this, we recall that Walsh functions $\Phi_{(m)}$ are related to block pulse functions $\Psi_{(m)}$ by the relation given by Equation 2.16.

Hence, the input waveform in Figure 4.12b can be represented in terms of block pulse functions simply by inspection and is given by

$$\mathbf{R1'}\ \Psi_{(m)} = [\underbrace{A\ A\ A\ \cdots\ A}_{n\ \text{terms}}\ Ax\ \underbrace{0\ 0\ 0\ \cdots\ 0}_{[(m/2)-n-1]\ \text{terms}}\ \underbrace{A\ A\ A\ \cdots\ A}_{n\ \text{terms}}\ Ax\ \underbrace{0\ 0\ 0\ \cdots\ 0}_{[(m/2)-n-1]\ \text{terms}}\ ]\Psi_{(m)} \qquad (4.41)$$

$$\underbrace{\hspace{10cm}}_{\text{Total } m \text{ terms}}$$

The above equation is a general expression for two repetition cycles. Suppose, for a particular input waveform of amplitude A, $n = 3$ and $m = 8$. Then

$$\mathbf{R2'}\ \Psi_{(m)} = [A\quad A\quad A\quad Ax\quad A\quad A\quad A\quad Ax]\ \Psi_{(8)} \qquad (4.42)$$

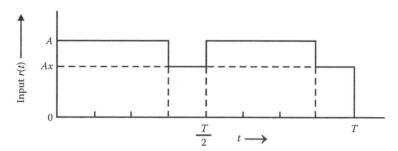

FIGURE 4.13
A typical pulse train for $n = 3$ and $m = 8$. (Reprinted from Deb, A. and Datta, A. K., *Int. J. Sys. Sci.*, 23, 151–166, 1992. With permission.)

This is represented in Figure 4.13.

By varying the number n and the magnitude of x in Equation 4.41, we can change the width of a pulse continuously.

For $n = m/2$, we have a full unchopped wave of amplitude A applied to the load. When $n = 0$, we have a pulse train having pulse width Ax and repetition cycle $T/2$ applied to the load. Thus, $0 \leq n \leq (m/2)$ and for different values of n, Equation 4.41 takes different modified forms.

Accommodating all these modifications, Equation 4.41 can now be written in the following matrix equation form:

$$\text{R3}' \; \Psi_{(m)} = \frac{m}{2} \text{rows} \left\{ \begin{bmatrix} Ax & 0 & 0 & \cdots & 0 & Ax & 0 & 0 & \cdots & 0 \\ A & Ax & 0 & \cdots & 0 & A & Ax & 0 & \cdots & 0 \\ A & A & Ax & \cdots & 0 & A & A & Ax & \cdots & 0 \\ A & A & A & \cdots & 0 & A & A & A & \cdots & 0 \\ \vdots & \vdots & \vdots & & \vdots & \vdots & \vdots & \vdots & & \vdots \\ A & A & \cdots & A & Ax & A & A & \cdots & A & Ax \end{bmatrix} \right. \Psi_{(m)} \quad (4.43)$$

$$\underbrace{\hphantom{Ax \; 0 \; 0 \; \cdots \; 0 \; Ax}}_{(m/2) \text{ columns}} \quad \underbrace{\hphantom{Ax \; 0 \; 0 \; \cdots \; 0 \; Ax}}_{(m/2) \text{ columns}}$$

The above equation represents all possible modes of PWM of an input waveform having a repetition cycle of $T/2$ seconds in a time period of T seconds.

Till now, we have not lifted the restriction on the repetition cycle and have considered only two pulses of the input waveform. However, this restriction could very well be removed by writing down a more general form of Equation 4.43 in which the repetition cycle is T_R seconds, and T/T_R number of pulses are considered in a period of T seconds. This is represented by the following equation:

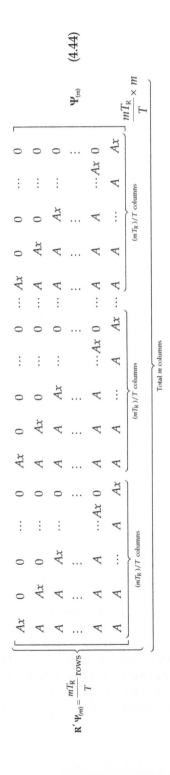

$$(4.44)$$

or

$$\mathbf{R}' \, \Psi_{(m)} = \mathbf{PWMM} \, \Psi_{(m)} \qquad (4.45)$$

We call the $(mT_R/T) \times (m)$ matrix on the right-hand side (RHS) of Equation 4.44 as PWM matrix (**PWMM**), which takes care of any possible PWM of the input waveform.

Equation 4.44 is the most generalized expression for any pulse-width modulated input $\mathbf{R}' \, \Psi_{(m)}$. For example, the nth row of Equation 4.44 will give that specific input in which each pulse covers n number of subintervals of T/m-second duration where its amplitude is A for first $(n-1)$ subintervals, Ax for the nth subinterval, and zero for the rest of $[(mT_R/T) - n]$ subintervals. This pulse profile is repeated T/T_R times over the total period of T seconds as is evident from Equation 4.44. By choosing proper values of m, T_R, x, and T, we can compute the output characteristics of the system for any duty interval as well as repetition rate.

4.3.2.1 Mathematical Operations

First of all, the pulse-width modulated input wave $\mathbf{R}' \, \Psi_{(m)}$ is converted to Walsh function domain by using Equation 2.16. Hence, from Equations 2.16 and 4.45, we have

$$\mathbf{R}\Phi_{(m)} = \{\mathbf{PWMM}\} \mathbf{W}^{-1} \, \Phi_{(m)} = \frac{1}{m} \{\mathbf{PWMM}\} \, \mathbf{W} \, \Phi_{(m)} \qquad (4.46)$$

Substituting $\mathbf{R}\Phi_{(m)}$ from the above equation in Equation 3.41, we have

$$\mathbf{C}\Phi_{(m)} = \frac{1}{m} [\{\mathbf{PWMM}\} \, \mathbf{W}] \, \{\mathbf{WOTF}\} \Phi_{(m)} \qquad (4.47)$$

To visualize the output easily, it is better to express $\mathbf{C}\Phi_{(m)}$ in terms of block pulse functions; hence, using Equation 2.16, we substitute $\Phi_{(m)}$ in the above equation to obtain

$$\mathbf{C}' \, \Psi_{(m)} = \frac{1}{m} \{\mathbf{PWMM}\} \mathbf{W} \, \{\mathbf{WOTF}\} \mathbf{W} \, \Psi_{(m)} \qquad (4.48)$$

where:
   $\mathbf{W}$ and $\mathbf{WOTF}$ are the square matrices of order m
   $\mathbf{PWMM}$ is of the dimension $[(mT_R/T) \times m]$

Hence, the matrix product on the RHS of the above equation gives a matrix of order $[(mT_R/T) \times m]$.
   Thus,

$$\mathbf{C}' \, \Psi_{(m)} = \frac{1}{m} \left[\text{product matrix} \right]_{\left(\frac{mT_R}{T} \times m\right)} \Psi_{(m)} \qquad (4.49)$$

$\mathbf{C}'\,\Psi_{(m)}$ gives the desired complete solution in the time domain since block pulse functions, like Walsh functions, are defined in the time domain. Evidently, each row of $\mathbf{R}'\,\Psi_{(m)}$ and $\mathbf{C}'\,\Psi_{(m)}$ corresponds, thus giving the solution for mT_R/T possible situations of PWM for any particular value of x in one mathematical operation.

4.3.2.2 Simulation of an Ideal Continuously Pulse-Width Modulated DC Chopper System

For ease of comparison, we consider the same load having $R_L = 31\ \Omega$ and $L_L = 0.17\ \text{H}$.

Considering the normalized load current I_L as the output variable and assuming that the output values are to be calculated in a time interval of 0.08 seconds, the **WOTF** in the scaled time domain is

$$\mathbf{WOTF3} = 31\left[31\,\mathbf{I} + \left(\frac{0.17}{0.08}\right)\mathbf{D}\right]^{-1} \tag{4.50}$$

where we choose $m = 16$. Hence, in the above equation, $\mathbf{I}$ is a unit matrix of order 16 and $\mathbf{D}$ is a square matrix of order (16×16).

By substituting the operational matrix $\mathbf{D}$ of order (16×16), as given by Chen et al. [22], the **WOTF3** could easily be evaluated.

We consider a pulse input having $T_R = 0.04$ seconds. Then, for such a pulse input of unit amplitude, the block pulse representation, considering all possible PWM strategies, as per Equation 4.43, may be given by

$$\mathbf{R}'\,\Psi_{(16)} = \begin{bmatrix} x & 0 & 0 & 0 & 0 & 0 & 0 & 0 & x & 0 & 0 & 0 & 0 & 0 & 0 & 0 \\ 1 & x & 0 & 0 & 0 & 0 & 0 & 0 & 1 & x & 0 & 0 & 0 & 0 & 0 & 0 \\ 1 & 1 & x & 0 & 0 & 0 & 0 & 0 & 1 & 1 & x & 0 & 0 & 0 & 0 & 0 \\ 1 & 1 & 1 & x & 0 & 0 & 0 & 0 & 1 & 1 & 1 & x & 0 & 0 & 0 & 0 \\ 1 & 1 & 1 & 1 & x & 0 & 0 & 0 & 1 & 1 & 1 & 1 & x & 0 & 0 & 0 \\ 1 & 1 & 1 & 1 & 1 & x & 0 & 0 & 1 & 1 & 1 & 1 & 1 & x & 0 & 0 \\ 1 & 1 & 1 & 1 & 1 & 1 & x & 0 & 1 & 1 & 1 & 1 & 1 & 1 & x & 0 \\ 1 & 1 & 1 & 1 & 1 & 1 & 1 & x & 1 & 1 & 1 & 1 & 1 & 1 & 1 & x \end{bmatrix}_{(8 \times 16)} \Psi_{(16)} \tag{4.51}$$

where:
$0 \le x \le 1$

Now knowing the Walsh matrix $\mathbf{W}$ of order (16×16), we can use Equation 4.48 to compute $\mathbf{C}'$, since **PWMM** is known from Equation 4.51 and **WOTF3** is known from Equation 4.50.

This is done with values of $x = 0.1, 0.2, \ldots, 0.9, 1.0$, which is equivalent to a modulation of the input pulse train by steps of 0.0005 seconds. If we like, we can

TABLE 4.3

Normalized Average Currents ($R_L = 31\ \Omega$, $L_L = 0.17$ H, $m = 16$)

Duty Cycle K_D	Normalized Load Current $I_{L(av)}$		Normalized Device Current $I_{s1(av)}$		Normalized Freewheeling Current $I_{F(av)}$	
	Exact	Walsh	Exact	Walsh	Exact	Walsh
0.1	0.0999	0.1000	0.0290	0.0251	0.0709	0.0749
0.2	0.1998	0.1999	0.0949	0.0856	0.1049	0.1143
0.25	0.2497	0.2498	0.1352	0.1322	0.1145	0.1176
0.3	0.2996	0.2997	0.1786	0.1681	0.1210	0.1316
0.4	0.3992	0.3994	0.2711	0.2635	0.1281	0.1358
0.5	0.4982	0.4987	0.3682	0.3669	0.1301	0.1318
0.6	0.5963	0.5969	0.4682	0.4610	0.1281	0.1358
0.7	0.6923	0.6928	0.5713	0.5612	0.1210	0.1316
0.75	0.7390	0.7404	0.6244	0.6228	0.1145	0.1176
0.8	0.7841	0.7840	0.6791	0.6697	0.1049	0.1143
0.9	0.8670	0.8658	0.7960	0.7909	0.0709	0.0749
1.0	0.9314	0.9315	0.9314	0.9315	0.0000	0.0000

choose even smaller variational steps for x. Naturally, with variation in x, K_D and K_F vary accordingly.

From the output stepped waveform $\mathbf{C}'\,\Psi$, the rms and average values of the normalized load current (I_L), the device current (I_{s1}), and the freewheeling current (I_F) are determined for different values of K_D. These are compared with respective values obtained from the exact equations given in Section 4.3.1.3 and are tabulated in Tables 4.3 and 4.4 (Program B.20 in Appendix B).

Table 4.3 compares the normalized average currents, whereas Table 4.4 compares the normalized rms currents. From these tables, it could be observed that the Walsh domain results are in good agreement with the exact solutions.

Comparing Tables 4.1 and 4.2 with Tables 4.3 and 4.4, respectively, it is noted that as expected, the accuracy of the computed results increases with an increase in m value. Tables 4.5 and 4.6 are formed to present a compact and clear picture of the reduction in percentage error of the average and rms load currents for three different values of K_D for increasing m from 8 to 16.

4.3.2.3 Determination of Normalized Average and rms Currents through Load and Semiconductor Components

As was done in Section 4.3.1.2, we proceed in a similar manner to determine the normalized average and rms currents through load and semiconductor components. The normalized average and rms load currents over the whole period are determined from Equation 4.48. The block pulse coefficients, say $c_0', c_1', \ldots, c_{14}', c_{15}'$, give the normalized average load current in each

TABLE 4.4

Normalized rms Currents $(R_L = 31\ \Omega,\ L_L = 0.17\ \text{H},\ m = 16)$

Duty Cycle K_D	Normalized Load Current $I_{L(rms)}$		Normalized Device Current $I_{s1(rms)}$		Normalized Freewheeling Current $I_{F(rms)}$	
	Exact	Walsh	Exact	Walsh	Exact	Walsh
0.1	0.1704	0.1583	0.1032	0.0709	0.1356	0.1324
0.2	0.3080	0.2925	0.2334	0.1718	0.2010	0.1820
0.25	0.3677	0.3636	0.2950	0.2854	0.2196	0.2253
0.3	0.4226	0.4100	0.3529	0.3030	0.2325	0.2134
0.4	0.5206	0.5133	0.4579	0.4312	0.2477	0.2377
0.5	0.6067	0.6057	0.5506	0.5486	0.2550	0.2567
0.6	0.6842	0.6790	0.6336	0.6092	0.2582	0.2361
0.7	0.7556	0.7489	0.7099	0.6803	0.2587	0.2294
0.75	0.7897	0.7888	0.7465	0.7451	0.2574	0.2586
0.8	0.8230	0.8172	0.7828	0.7589	0.2540	0.2291
0.9	0.8877	0.8845	0.8577	0.8456	0.2291	0.2118
1.0	0.9472	0.9472	0.9472	0.9472	0.0000	0.0000

TABLE 4.5

Reduction in Percentage Error in Computation of Normalized Average Load Current for Increasing m from 8 to 16

Duty Cycle K_D	Percentage Error in Normalized Average Load Current $I_{L(av)}$		Reduction in Absolute Value of Percentage Error for Increasing m from 8 to 16
	Walsh Method $m = 8$	Walsh Method $m = 16$	
0.25	−0.120	−0.040	0.080
0.50	−0.341	−0.100	0.241
0.75	−1.083	−0.189	0.894

TABLE 4.6

Reduction in Percentage Error in Computation of Normalized rms Load Current for Increasing m from 8 to 16

Duty Cycle K_D	Percentage Error in Normalized rms Load Current $I_{L(rms)}$		Reduction in Absolute Value of Percentage Error for Increasing m from 8 to 16
	Walsh Method $m = 8$	Walsh Method $m = 16$	
0.25	6.092	1.115	4.977
0.50	0.659	0.165	0.494
0.75	0.608	0.114	0.494

consecutive subinterval of 0.005 seconds starting from 0. Hence, in line with Equations 4.22 and 4.23, we can write

$$I_{L(av)} = \mathbf{C}' \, \Psi_{(16)}(t)_{(av)} = \frac{1}{16} \sum_{n=0}^{15} c_n' \tag{4.52}$$

and

$$I_{L(rms)} = \mathbf{C}' \, \Psi_{(16)}(t)_{(rms)} = \sqrt{\left[\frac{1}{16} \sum_{n=0}^{15} c_n'^{\,2} \right]} \tag{4.53}$$

However, it should be kept in mind that the above equations apply for each row of the coefficient matrix of the output Equation 4.49. For our case, there will be eight rows for eight different situations of PWM.

Similar to Equations 4.24 and 4.25, the average and rms values of the normalized device currents over the whole period of 0.08 seconds are given by

$$I_{s1(av)} = \frac{1}{16} \sum_{n=0}^{15} c_n' \delta \tag{4.54}$$

and

$$I_{s1(rms)} = \sqrt{\left[\frac{1}{16} \sum_{n=0}^{15} c_n'^{\,2} \delta^2 \right]} \tag{4.55}$$

Also, we express the average and rms values of the normalized freewheeling current as

$$I_{F(av)} = \frac{1}{16} \sum_{n=0}^{15} c_n' \bar{\delta} \tag{4.56}$$

and

$$I_{F(rms)} = \sqrt{\left[\frac{1}{16} \sum_{n=0}^{15} c_n'^{\,2} \bar{\delta}^2 \right]} \tag{4.57}$$

where:
δ is the logic operator defined in Section 4.3.1.2

While the use of Equations 4.52 and 4.53 are straightforward, we find that this is not the case for Equations 4.54 through 4.57. This is because of the fact that the division of the Walsh domain output load current (I_L) into two parts, namely, device current (I_{s1}) and freewheeling current (I_F), is not straightforward like the previous case of stepwise PWM. Since we consider every possible situation of PWM, it is possible that the trailing edges of the input pulses may not coincide with any two subinterval interface. Thus, one input pulse may not cover the integral number of subintervals, each of T/m-second duration. This has been explained in detail in Section 4.3.2, and to apply the Walsh domain method of analysis, we have employed the area-preserving technique to convert an "irregular" input pulse waveform into a "regular" one (Figure 4.12).

Now to divide the Walsh domain output load current into two parts, for example, device current and freewheeling current—we again apply the area-preserving technique as and when necessary. Referring to Figure 4.14, let us suppose that the trailing edge of any input pulse lies in the nth subinterval (i.e., end of time $K_D T_R$ and beginning of time $K_F T_R$). Then, the output block of amplitude N (say) of $\mathbf{C}' \Psi(t)$ lying in the same nth subinterval is divided into two parts, namely, xN and $(1 - x) N$, by a hypothetical vertical line. This is done in a manner already outlined in Section 4.3.2. Now, xN is considered to be a part of device current and is used for computation of the same. Similarly, $(1 - x)N$ is used for freewheeling current determination. Thus, xN and $(1 - x)N$ are among the coefficients c'_n of the output block pulse functions.

Hence, Equations 4.52 through 4.57 can now be applied for the determination of the normalized average and rms currents, even though we have an "irregular" input pulse waveform at our disposal.

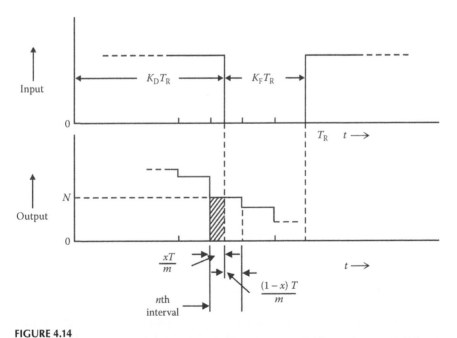

FIGURE 4.14

Separation technique for the main thyristor current (I_{s1}) and the freewheeling current (I_F) when the trailing edge of the input pulses does not coincide with any two subinterval interfaces. (Reprinted from Deb, A. and Datta, A. K., *Int. J. Sys. Sci.*, 23, 151–166, 1992. With permission.)

For a comparison of the output load current waveforms with those obtained from exact solutions, Figure 4.15 is drawn for three different values of the duty cycle K_D (=0.25, 0.50, and 0.75).

Comparing Figure 4.15 with Figure 4.10, it is found that there has been a marked improvement in the staircase approximation of the exact waveforms by increasing m from 8 to 16.

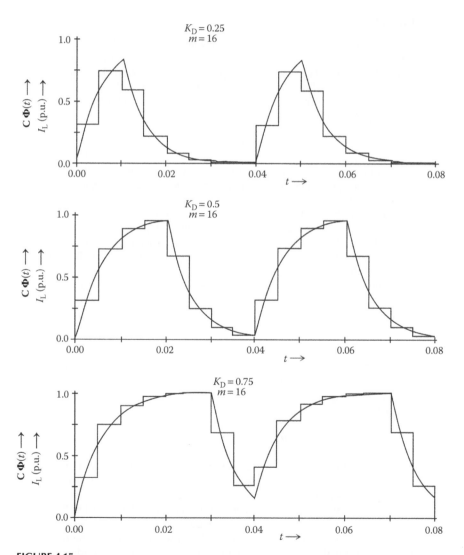

FIGURE 4.15
Exact and Walsh/block pulse waveforms of the normalized load current for $m = 16$, $T = 0.08$ seconds, and $K_D = 0.25$, 0.50, and 0.75. (Reprinted from Deb, A. and Datta, A. K., *Int. J. Sys. Sci.*, 23, 151–166, 1992. With permission.)

From Tables 4.3 and 4.4, it is observed that compared to the load current (I_L) and the device current (I_{s1}), there is a slightly more deviation of the freewheeling current (I_F) from the predicted behavior. It may be due to an "irregular" input waveform for which the trailing edge of the pulse does not coincide with any two subinterval interfaces, and it should also be noted that the variable I_F is not of monotonically rising nature like the load current or device current.

Example 4.1

Consider a chopper circuit with a supply voltage of $E = 100$ V. The inductive and resistive loads are $R_L = 2.2\ \Omega$ and $L_L = 0.0002$ H for continuous conduction. The duty cycle of the chopper is $K_D = 0.75$. Also, the chopper is "on" for 50 μs and remains in the "off" state for 16.67 μs [23].

Within the time range $0 < t < T_R$, the input voltage is defined in terms of block pulse functions as

$$r(t) \approx [100\ 100\ 100\ 0]\Psi_{(8)}(t)$$

Now, following the mathematical steps described earlier, we find the rms and average values of the load current, the device current, and the diode current.

Here, since the problem is about continuous conduction, m is considered large for better accuracy. Also, the output is determined for a wide range of time: $0 < T < 0.0043$ seconds. For better clarity, the output waveform is shown for $0 < T < 0.0015$ seconds in Figure 4.16 (Program B.21 in Appendix B).

All the parameters are shown in Table 4.7 for a comparative study.

Example 4.2

Consider a chopper circuit with a supply voltage of $E = 220$ V. The inductive and resistive loads are $R_L = 5\ \Omega$ and $L_L = 7.5$ mH for continuous conduction. The duty cycle of the chopper is $K_D = 0.50$. Also, the chopper frequency is given as 1 kHz [24].

Within the time range $0 < t < T_R$, the input voltage is defined in terms of block pulse functions as

$$r(t) \approx [220\ 220\ 0\ 0]\Psi_{(8)}(t)$$

Now, proceeding in a similar manner, we determine the rms and average values of the load current, the device current, and the diode current. These parameters are summarized in Table 4.8 for a comparative study. Again, for continuous conduction, m is considered to be large, that is, 256. Also, the output waveform is shown in Figure 4.17, which resembles the exact solution given in Ref. [24]. As in the previous example, the output is determined for a wide range of time: $0 < T < 0.0640$ seconds. For better clarity, the output waveform is plotted for $0 < T < 0.02$ seconds (Program B.22 in Appendix B).

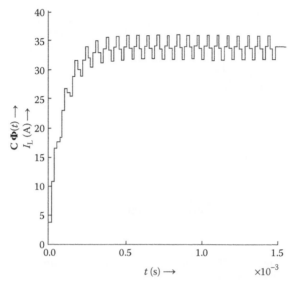

FIGURE 4.16
Walsh domain solution ($m = 256$, $T = 0.0043$ seconds) of the load current of Example 4.1. The output is shown for 0.0015 seconds for better clarity.

TABLE 4.7

Comparison of Different Parameters

	Average Value		rms Value	
Variable	Exact Values from Ref. [23]	Walsh Method	Exact Values from Ref. [23]	Walsh Method
Load current (A)	–	33.4348	34.15	33.6148
Device current (A)	–	25.0813	29.53	29.1471
Freewheeling diode current (A)	–	8.3535	17.05	16.7452

TABLE 4.8

Comparison of Different Parameters

	Average Value		rms Value	
Variable	Exact Values from Ref. [24]	Walsh Method	Exact Values from Ref. [24]	Walsh Method
Load current (A)	22	21.5697	22.10	21.7337
Device current (A)	–	10.8251	15.63	15.4462
Freewheeling diode current (A)	–	10.7446	–	15.2895

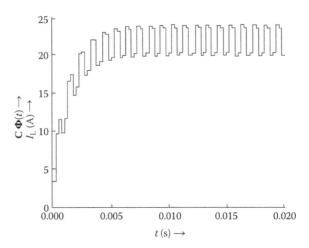

FIGURE 4.17
Walsh domain solution ($m = 256$, $T = 0.0640$ seconds) of the load current of Example 4.2. The output is shown for 0.02 seconds for better clarity.

4.3.2.4 Simulation of an Ideal Chopper-Fed DC Series Motor

In a series motor, the field winding is connected in series with the armature. For a chopper-fed series motor, the motor terminal voltage is controlled by a first-quadrant chopper, and the schematic diagram of such a motor drive system is shown in Figure 4.18 [16,17,25]. It should be noted that when the main thyristor s1 is turned off, the load current i_a flows through the diode D_F connected across the motor load.

We consider the thyristor as an ideal switch and neglect the switching time of the semiconductor devices as these are small compared to the usual pulse time. The underlying assumptions for such a motor drive system are as follows [26]:

1. During a certain pulse condition, the speed N of the motor remains unchanged.
2. The inductance of the armature circuit remains constant. Actually, this value decreases somewhat with increasing current and flux due to saturation.
3. The effect of circulating current in the commutating coils is neglected.
4. Operation of the motor is in the linear region of the magnetization curve.
5. The brushes are located on the magnetic neutral axis, and all armature feedback effects are neglected.
6. Voltage drops in brushes or semiconductor devices are neglected.

Type	Chopper configuration	$E_o - I_o$ Characteristics	Function
First-quadrant chopper	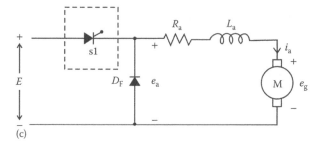		$e_o = E$, s1 on $= 0$, s1 off D_F on

(a)

(b)

(c)

FIGURE 4.18
(a) Chopper configuration, (b) schematic diagram of a chopper-fed DC series motor, and (c) overall schematic diagram.

Now the basic motor equations, as found in the literature, are presented below.

The armature circuit resistance R_a and inductance L_a include the resistance and inductance of the series field winding, respectively, as well. The back electromotive force (emf) is

$$e_g = K_a \phi n \tag{4.58}$$

The flux has two components. One component, say ϕ_a, is produced by the armature current flowing through the series field winding. The other component, say ϕ_{res}, is due to the so-called residual magnetism. The latter is small and can be assumed constant.

Hence,

$$\phi = \phi_a + \phi_{res} \tag{4.59}$$

As magnetic linearity is assumed, we can write

$$\phi_a = K_f i_a \tag{4.60}$$

From the above equations, we have

$$e_g = K_{af} i_a n + K_{res} n \tag{4.61}$$

where:
  n is the instantaneous speed of the motor
  K_a, K_f, and K_{res} are DC machine constants with $K_{af} = K_a K_f$

The contribution of residual magnetism in generating the back emf is very small and is proportional to speed, while the major contribution comes from the flux produced by the armature current. Now, the average back emf is given by

$$E_g = K_{af} I_a N + K_{res} N \tag{4.62}$$

The developed instantaneous torque is

$$\tau = K_a \phi i_a \tag{4.63}$$

Neglecting residual magnetism, we have

$$\tau \approx K_{af} i_a^2 \tag{4.64}$$

Hence, the average toque is

$$\tau \approx K_{af} I_{ar}^2 \tag{4.65}$$

where:
  I_{ar} is the rms motor current

The armature circuit voltage equation is

$$e_a = R_a i_a + L_a \frac{di_a}{dt} + e_g \tag{4.66}$$

In terms of average quantities,

$$E_a = R_a I_a + E_g \tag{4.67}$$

The basic DC motor Equations 4.62, 4.65, and 4.67 hold good for the chopper-controlled DC series motor drive in terms of average quantities.

As mentioned in Sections 3.3 and 4.3.2, the present method requires a linearized model for handling a nonlinear load. Hence, we use such a model for simulation of a DC series motor [25,27] and find out the responses for average load current and speed for small voltage perturbation around a stable operating point. A linear model helps us to define a transfer function as well as a **WOTF** to make use of Equation 3.41.

4.3.2.4.1 Linearized Model

A linear model or transfer function for the series motor can be derived by applying small-signal perturbation technique. This model can then be used in the analysis of a closed-loop system incorporating the series motor. The analysis is valid for small disturbances around a given steady-state operating point.

The differential equations governing the performance of the DC series motor are, from Equations 4.61, 4.64, and 4.66, given by

$$e_a = R_a i_a + L_a \frac{di_a}{dt} + K_{af} i_a n + K_{res} n \tag{4.68}$$

$$K_{af} i_a^2 = J \frac{dn}{dt} + Bn + \tau_L \tag{4.69}$$

where:
  J is the moment of inertia
  B is the coefficient of viscous friction
  τ_L is the load torque

These equations are valid for both transient and steady-state conditions. They are nonlinear because of the product and square terms involving the variables. In addition, variation of the parameters, such as R_a, L_a, K_{af}, and B, makes the equations nonlinear. To derive a linearized model or transfer function, the following assumptions are made:

1. The parameters remain constant for a small perturbation.
2. During the transient period after the disturbance, the transient behavior is considered in terms of time variation of average values of current and speed, rather than their instantaneous values, that is, the system passes through quasi-steady-state operating conditions during the disturbance.

Since the current is unidirectional, the average voltage across the inductance is zero over a cycle. In terms of average values, the defining equations are

$$E_a = I_a R_a + K_{af} I_a N + K_{res} N \tag{4.70}$$

$$K_{af} I_a^2 = J \frac{dn}{dt} + BN + \tau_L \tag{4.71}$$

Small-signal linearization around an operating point gives

$$\Delta E_a = \Delta I_a (R_a + K_{af}N_o) + \Delta N (K_{res} + K_{af}I_{ao}) \tag{4.72}$$

$$2K_{af}I_{ao}\,\Delta I_a = Jp\,\Delta N + B\Delta N + \Delta\tau_L \tag{4.73}$$

where:
$p = d/dt$
I_{ao} is the motor current at the operating point
N_o is the motor speed at the operating point

In the Laplace domain, these equations can be rewritten as follows:

$$\Delta I_a(s) = \frac{\Delta E_a(s) - \Delta N(s)(K_{res} + K_{af}I_{ao})}{R_a + K_{af}N_o} \tag{4.74}$$

$$\Delta N(s) = \frac{2K_{af}I_{ao}\,\Delta I_a(s) - \Delta\tau_L(s)}{B + Js} \tag{4.75}$$

These equations in the block diagram representation are shown in Figure 4.19a.
However, if we consider the response due to changes in motor voltage only, the block diagram simplifies to that of Figure 4.19b because $\Delta\tau_L(s) = 0$. The transfer function relevant to speed changes is given by

$$\frac{\Delta N(s)}{\Delta E_a(s)} = \frac{G_E(s)}{1 + G_E(s)H_E(s)} \tag{4.76}$$

where:

$$G_E(s) = \left(\frac{2K_{af}I_{ao}}{R_a + K_{af}N_o} \right) \left(\frac{1}{B + Js} \right) \tag{4.77}$$

and

$$H_E(s) = K_{res} + K_{af}I_{ao} \tag{4.78}$$

From the Equations 4.76 through 4.78,

$$\frac{\Delta N(s)}{\Delta E_a(s)} = \frac{2K_{af}I_{ao}}{R_a + K_{af}N_o} \cdot \frac{1}{B + \dfrac{2K_{af}I_{ao}(K_{res} + K_{af}I_{ao})}{(R_a + K_{af}N_o)} + Js} \tag{4.79}$$

$$= k_{EN}\frac{1}{1 + sT_o}$$

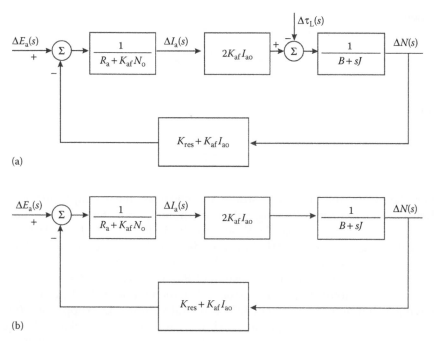

FIGURE 4.19
(a) Block diagram representation of a linearized model of DC series motor and (b) modified block diagram for small-signal perturbation around a stable operating point.

where:

$$T_o = \frac{J}{B_o}$$

$$B_o = B + \frac{2K_{af}I_{ao}(K_{res}+K_{af}I_{ao})}{(R_a + K_{af}N_o)}$$

and

$$k_{EN} = \frac{2K_{af}I_{ao}}{B_o(R_a + K_{af}N_o)}$$

The transfer function for current changes is given by

$$\frac{\Delta I_a(s)}{\Delta E_a(s)} = \frac{\Delta I_a(s)}{\Delta N(s)} \cdot \frac{\Delta N(s)}{\Delta E_a(s)}$$

$$= \frac{B+Js}{2K_{af}I_{ao}} k_{EN} \frac{1}{1+sT_o} \qquad (4.80)$$

$$= k_{E1} \frac{1+sT_m}{1+sT_o}$$

where:

$$k_{E1} = \frac{B}{2K_{af}I_{ao}} k_{EN}$$

and

$$T_m = \frac{J}{B}$$

For the series motor model study, we consider the example given in Ref. [25] having the following values of motor parameters and constant: $K_{af} = 0.027$ H, $K_{res} = 0.0273$ V seconds/rad, $R_a = 1$ Ω, $L_a = 0.032$ H, $J = 0.0465$ kg – m², $B = 0.004$ N m seconds/rad, $E_a = 102.62$ V, $\Delta E_a = 5.4$ V (=5% of E_a), $N_o = 1485$ rpm, $I_{ao} = 18.72$ A, and $\tau_L = 9$ N m.

In order to compare with the results of Ref. [25], we consider an average change in the motor supply voltage of 5.4 V, which is obtainable by a chopper having a chopping frequency of 125 ms operated from a DC supply of 120 V. A change of duty cycle K_D from 0.855 to 0.9 will achieve a voltage change from 102.6 to 108 V, which is equivalent to a change of average voltage of 5.4 V by phase-control technique [25].

Substituting the values of the motor parameters and constants in Equations 4.79 and 4.80, we have

$$\frac{\Delta N(s)}{\Delta E_a(s)} = \frac{1.8073}{(1+0.4322s)} \tag{4.81}$$

and

$$\frac{\Delta I_a(s)}{\Delta E_a(s)} = \frac{0.0071514(1+11.625s)}{(1+0.4322s)} \tag{4.82}$$

We consider a time interval of $T = 2.0$ seconds, and hence the scaling constant $\lambda = 2$. Now following the operational technique outlined in Chapter 3 and Section 4.3.2.1, we use Equations 4.81 and 4.82 to obtain the respective WOTFs and consequently the Walsh domain solutions for the transient response of speed and current due to a step change in motor supply voltage, while load torque is kept constant. Here, we use 16 basis Walsh functions for the computations.

The results obtained by sequence domain analysis are plotted in Figure 4.20 and compared with the response obtained by conventional methods. A good agreement is readily observed.

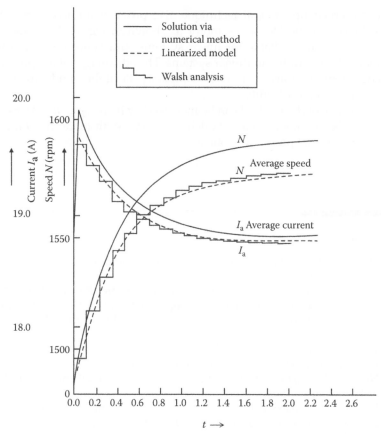

FIGURE 4.20
Comparison of solutions for average speed and average current of a DC series motor obtained by different methods of analysis. For Walsh analysis, we have considered $m = 16$ and $T = 2$ seconds. (Reprinted from Deb, A. and Datta, A. K., *Int. J. Sys. Sci.*, 23, 151–166, 1992. With permission.)

4.4 Conclusion

In this chapter, we have presented in detail how the proposed sequency domain technique could be applied successfully to pulse-fed single-input single-output (SISO) systems and pulse-width modulated PE systems. It was also shown how the accuracy of the Walsh domain analysis is enhanced when 16 basis functions are used instead of 8. This enhancement is evident from Tables 4.5 and 4.6, and also from Figure 4.15. Tables 4.3 and 4.4 show that the continuous PWM technique in the Walsh domain gives results well within tolerable error limit.

The presented operational technique also provides an acceptable staircase solution for the DC series motor as shown in Figure 4.20. However, the method is only suitable for linear systems, and hence needs a linearized model for handling nonlinear systems. The advantages of the presented sequency domain analysis lie in its simplicity, flexibility, and direct computer or digital compatibility.

The accuracy of the method can be increased by increasing $m = 32, 64, 128$, or even higher number of basis Walsh functions. With modern computers, this surely does not pose any problem.

References

1. Dewan, S. B. and Straughen, A., *Power Semiconductor Circuits*, John Wiley & Sons, New York, 1975.
2. Datta, A. K., Deb, A., Roy, A., Roychowdhury, A. and Roy, B., R and D and operational experience of thyristor chopper controller for Calcutta tram, IEEE Conference on Power Electronics and Variable Speed Drives, London, May 1–4, 1984, pp. 230–232.
3. Takeuchi, T. J., *Theory of SCR Circuit and Application to Motor Control*, Tokyo Electrical Engineering College Press, Tokyo, 1968.
4. Nixon, F. E., *Handbook of Laplace Transformation* (2nd Ed.), Prentice Hall, Upper Saddle River, NJ, 1965.
5. Jacovides, L., Analysis of induction motor drives with a nonsinusoidal supply voltage using Fourier analysis, *IEEE Trans. Ind. Appl.*, Vol. **IA-9**, No. 4, pp. 741–747, 1973.
6. Hoft, R. G. and Casteel, J. B., Power electronic circuit analysis techniques, Proceedings of the IFAC Symposium on Control in Power Electronics and Electrical Drives, Dusseldorf, Germany, October 1977, pp. 987–1024.
7. Nayak, P. H. and Hoft, R. G., Computer-aided steady-state analysis of thyristor DC drives on weak power systems, IEEE Conference on Record of 11th Annual Meeting Industry Application Society, 1976, pp. 835–847.
8. Ramamoorty, M. and Ilango, B., Application of steady-space techniques to the steady-state analysis of thyristor controlled single-phase induction motors, *Int. J. Control*, Vol. **16**, No. 3, pp. 353–368, 1972.
9. Novotony, D. W., Switching function representation of polyphase inverters, IEEE Conference on Record of 10th Annual Meeting Industry Application Society, Atlanta, GA, September 28–October 2, 1975, pp. 823–831.
10. Gyugyi, L. and Pelly, B. R., *Static Power Frequency Changers*, Wiley Intersciences, New York, 1976.
11. Wood, P., *Switching Power Converters*, Van Nostrand Reinhold, New York, 1981.
12. Parrish, E. A. and McVey, E. S., A theoretical model for single-phase silicon controlled rectifier systems, *IEEE Trans. Automat. Contr.*, Vol. **AC-12**, No. 5, pp. 577–579, 1979.

13. Sucena-Paiva, J. P. et al., Stability study of controlled rectifiers using a new discreate model, *Proc. IEE*, Vol. **119**, No. 10, pp. 1285–1293, 1972.
14. Harmuth, H. F., *Transmission of Information by Orthogonal Functions* (2nd Ed.), Springer-Verlag, Berlin, 1972.
15. Deb, A. and Datta, A. K., Time response of pulse-fed SISO systems using Walsh operational matrices, *Adv. Model. Simul.*, Vol. **8**, No. 2, pp. 30–37, 1987.
16. Deb, A., Sarkar, G., Sen, S. K. and Dutta, A. K., A new method of analysis of chopper fed DC series motor using Walsh/block pulse function, *Ind. J. Eng. Mater. Sci.*, Vol. **6**, pp. 330–334, 1999.
17. Deb, A. and Datta, A. K., A new method of analysis of chopper-fed DC series motor using Walsh function, Proceedings of the 4th European Conference on Power Electronics and Applications (EPE '91), Florence, September, 3–6, 1991.
18. Deb, A. and Datta, A. K., Analysis of pulse-fed power electronic circuits using Walsh functions, *Int. J. Electron.*, Vol. **62**, No. 3, pp. 449–459, 1987.
19. Deb, A. and Datta, A. K., On analytical techniques of power electronic circuit analysis. *IETE Tech. Rev.*, Vol. **7**, No. 1, pp. 25–32, 1990.
20. Deb, A. and Datta, A. K., Analysis of a continuously variable pulse-width modulated system via Walsh functions, *Int. J. Sys. Sci.*, Vol. **23**, No. 2, pp. 151–166, 1992.
21. Revankar, G. N. and Tandon, V. K., Ideal choppers and pulse-width modulated converter systems, *IEEE Trans. Industr. Electron. Control Instrum.*, Vol. **IECI-22**, No. 3, pp. 417–420, 1975.
22. Chen, C. F., Tsay, T. Y. and Wu, T. T., Walsh operational matrices for fractional calculus, *J. Franklin Inst.*, Vol. **303**, No. 3, pp. 267–284, 1977.
23. Shepherd, W. and Zhang, L., *Power Converter Circuits*, Taylor & Francis, New York, 2010.
24. Rashid, M. H., *Power Electronics Circuits, Devices, and Applications* (3rd Ed.), Pearson, New Delhi, 2004.
25. Sen, P. C., *Thyristor DC Drives*, John Willey & Sons, New York, 1981.
26. Franklin, P. W., Theory of the DC motor controlled by power-pulses: Part I—motor operation, IEEE Conference on Record of 5th Annual Meetings Industry and General Applications Group, pp. 59–67, 1970.
27. Ramamoorty, M. and Ilango, B., The transient response of a thyristor-controlled series motor, *IEEE Trans. Power Appar Syst.*, Vol. **90**, pp. 289–297, 1971.

5

Analysis of Controlled Rectifier Circuits

Traditionally, conversion of AC to DC voltages has been dominated by phase-controlled or diode rectifiers. Three-phase fully controlled bridge rectifiers are well suited for several power electronic (PE) applications that require a large current such as brushless excitation systems for aircraft generators, high-voltage AC-to-DC converters, and DC motor speed control systems [1–4]. The performance of a slip-ring induction motor can be controlled by a silicon-controlled rectifier (SCR) [5].

Different technologies [6] are used in the generation of large controlled currents in the kiloampere range. Microcontroller- and microprocessor-based control techniques and design of three-phase and single-phase controlled rectifiers have also been proposed [7,8].

The ideal character of the input currents drawn by these rectifiers creates a number of problems such as increase in reactive power, high-input current harmonics and low-input power factor, lower rectifier efficiency, and large input voltage distortion etc. Power factor has been improved in various ways [7,9,10] for single-phase controlled rectifiers. Harmonics are analyzed using the bridge configuration for controlled rectifier circuits [11].

A new method [12] has been developed to design a single-phase rectifier with improved power factor and low total harmonic distortion (THD) using boost converter technique.

Phase-controlled rectifier circuits and PE circuits have also been analyzed in Walsh and block pulse domain [13,14].

In general, controlled rectifiers are fed from AC sources of constant frequency. The current through the load depends on the triggering (firing) angle α of the controlled rectifier and the phase angle φ of the connected load. We will consider in the following a half-wave as well as a full-wave controlled rectifier, fed from fixed-frequency sinusoidal supply, connected to an inductive load. From the Walsh domain analysis, variations of the normalized average and root mean square (rms) load currents with the firing angle α for a half-wave controlled rectifier are plotted and compared with the curves obtained from traditional analysis. For the full-wave rectifier, comparison tables are formed for the normalized average load currents for different firing angles.

The basic approach of the above-mentioned analysis is the same as that presented in Chapters 2 and 3. The main steps are as follows:

1. To represent the input load voltage waveform in terms of Walsh functions. Let this be $\mathbf{R}\Phi(t)$.

2. To represent the load system by a Walsh operational transfer function (**WOTF**) by replacing the Laplace operator s in its transfer function by the Walsh domain operational matrix for differentiation **D**

3. To find the matrix product $[\mathbf{R}\ \mathbf{WOTF}\ \Phi(t)]$, which will give the output response $\mathbf{C}\Phi(t)$ in the Walsh domain. Since Walsh functions are defined in time domain, $\mathbf{C}\Phi(t)$ will give the time response of the system.

5.1 Representation of a Sine Wave by Walsh Functions

As discussed in Chapter 3, a sine wave could very well be represented by a set of Walsh functions. We use Equations 3.1 and 3.2 to find out the coefficients c_0, c_1,... of different Walsh functions, which form the desired sine wave when combined together. We consider a sine wave of frequency f cycles per second and concentrate our attention on a single-cycle operation when the circuit is initiated at $t = 0$.

Following the method outlined in Section 3.2 and considering the first eight basis Walsh functions, the Walsh series coefficients of the sine wave cycle in the scaled domain can be expressed as

$$c_{0\lambda} = \frac{1}{\lambda} \int_0^\lambda \phi_0(\lambda t) \cdot \sin(2\pi f t) dt$$

$$c_{1\lambda} = \frac{1}{\lambda} \int_0^\lambda \phi_1(\lambda t) \cdot \sin(2\pi f t) dt$$

$$c_{2\lambda} = \frac{1}{\lambda} \int_0^\lambda \phi_2(\lambda t) \cdot \sin(2\pi f t) dt$$

$$\vdots$$

$$c_{7\lambda} = \frac{1}{\lambda} \int_0^\lambda \phi_7(\lambda t) \cdot \sin(2\pi f t) dt$$

where:

$c_{0\lambda}$, $c_{1\lambda}$,... are the coefficients of respective Walsh functions in the scaled Walsh domain

λ is the scaling coefficient given by $\lambda = 1/f$.

Evaluation of integrals within proper limits yields the values of the coefficients, and the sine wave is given by

$$\sin(2\pi ft) \approx [0 \quad 0.637 \quad 0 \quad 0 \quad 0 \quad 0 \quad 0 \quad -0.264]\Phi_{(8)}(\lambda t) \quad (5.1)$$

Transforming Equation 5.1 to block pulse domain, the same sine waveform is represented as

$$\sin(2\pi ft) \approx [0.373 \; 0.901 \; 0.901 \; 0.373 \; -0.373 \; -0.901 \; -0.901 \; -0.373] \, \Psi_{(8)}(\lambda t) \; (5.2)$$

From Equation 5.2, we find that the average value of the stepped waveform is 0.637, which is equal to the normalized average value of a sinusoid: $(2/\pi) \approx 0.6366$. Figure 5.1a shows the actual waveform along with its Walsh function representation considering the eight basis functions.

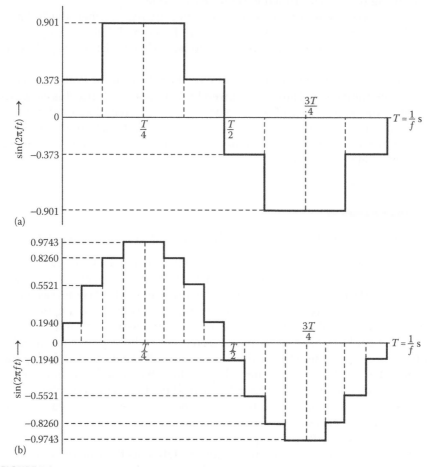

FIGURE 5.1
Representation of a sine wave by Walsh/block pulse functions (a) using 8 basis functions and (b) using 16 basis functions. (Reprinted from Deb, A. and Datta, A. K., *Int. J. Electron.*, 79, 861–883, 1995. With permission.)

If we consider 16 basis Walsh functions, the same procedure yields the following set of coefficients that, when combined, form the resulting sine wave:

$$\sin(2\pi f t) \approx [0 \quad 0.6366 \quad 0 \quad 0 \quad 0 \quad 0 \quad 0 \quad -0.2636 \quad 0$$
$$0 \quad 0 \quad -0.1266 \quad 0 \quad -0.0524 \quad 0 \quad 0]\Phi_{(16)}(\lambda t) \tag{5.3}$$

In terms of block pulse functions, the above equation could be expressed as

$$\sin(2\pi f t) \approx [0.1940 \; 0.5521 \; 0.8260 \; 0.9743 \; 0.9743 \; 0.8260 \; 0.5521 \; 0.1940 \; -0.1940$$
$$-0.5521 \; -0.8260 \; -0.9743 \; -0.9743 \; -0.8260 \; -0.5521 \; -0.1940]\Psi_{(16)}(\lambda t) \tag{5.4}$$

Figure 5.1b shows the Walsh or block pulse function representation of a sine wave using the 16 basis function set for a full cycle.

As expected, the accuracy of representation increases when the 16 basis Walsh functions are used instead of 8. From Equation 5.4, the average value of the transformed sine wave turns out to be 0.6366, which is very close to the actual value of $(2/\pi)$.

In presenting the controlled rectifier circuit analysis in the Walsh domain, we use the transformed sine wave given by Equation 5.3. Since it is easier to visualize the stepped waveform in block pulse representation, we deal with the input sine wave in block pulse domain by using Equation 5.4 and then transform it back to the Walsh domain using Equation 2.16 for subsequent mathematical operations.

5.2 Conventional Analysis of Half-Wave Controlled Rectifier

Half-wave controlled rectifier circuits are well known and so are their application areas [15–17]. We consider a controlled half-wave rectifier, supplied from an ideal sinusoidal voltage source, feeding a linear load comprised of an inductance L_L and a resistance R_L. The arrangement is shown in Figure 5.2a.

Let us assume that the thyristor s1 behaves as an ideal switch and is turned "on" at an angle α from the leading zero crossing of the impressed sinusoidal voltage $v(t) = \sqrt{2}V \sin(\omega t)$. The thyristor is naturally commutated during the negative half-cycle of the forcing voltage, where the period of commutation depends on the load time constant. In the next positive half-cycle, the thyristor is again turned "on" at an angle α and the first cycle is repeated periodically keeping α constant.

To determine the forced response due to such a nonsinusoidal periodic excitation, we traditionally represent the excitation as the sum of an infinite series of sinusoidal functions of integral multiple of the source frequency f. The response due to each significant term of the series is then determined.

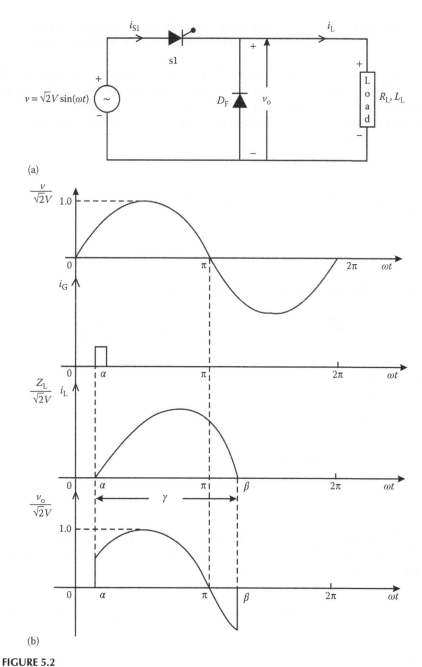

(a)

(b)

FIGURE 5.2
(a) Controlled half-wave rectifier feeding an RL load and (b) typical waveforms of normalized impressed voltage, gate signal, normalized load current, and load voltage. (Reprinted from Deb, A. and Datta, A. K., *Int. J. Electron.*, 79, 861–883, 1995. With permission.)

Since the load is linear, the forced responses due to all significant terms may be combined by superposition. In short, we employ the well-known Fourier analysis.

Figure 5.2b shows the typical waveforms of the gate signal, the normalized load voltage ($v_o/\sqrt{2V}$), and the normalized load current ($i_L Z_L/\sqrt{2V}$) along with the normalized impressed voltage ($v/\sqrt{2V}$).

In Figure 5.2b, γ is the conduction angle and β is the extinction angle having the relation:

$$\beta = \gamma + \alpha$$

When thyristor s1 is gated, the circuit current i_L is given by the following differential equation:

$$L_L \frac{di_L}{dt} + R_L i_L = \sqrt{2}V \sin(\omega t) \tag{5.5}$$

Solution of the above equation is obtained as [18]

$$i_L = \left(\frac{\sqrt{2}V}{Z_L} \right) \left[\sin(\omega t - \varphi) - \sin(\alpha - \varphi) \exp\left\{ \left(\frac{R_L}{L_L} \right)\left(\frac{\alpha}{\omega} - t \right) \right\} \right] \tag{5.6}$$

where:
$$Z_L = \sqrt{\left[R_L^2 + \omega^2 L_L^2 \right]}$$
$$\varphi = \tan^{-1}(\omega L_L / R_L)$$

At $\omega t = \beta$ and $i_L = 0$, from Equation 5.6 we have

$$\sin(\beta - \varphi) = \sin(\alpha - \varphi) \exp\left\{ \frac{R_L(\alpha - \beta)}{\omega L_L} \right\} \tag{5.7}$$

From the transcendental above equation, β may be solved numerically.

To obtain the normalized value of the average and rms rectified currents using Equation 5.6, these are given by

$$I_{L(av)} = \frac{1}{2\pi} \int_{\alpha}^{\alpha+\gamma} \left[\sin(\omega t - \varphi) - \sin(\alpha - \varphi) \exp\left\{ \left(\frac{R_L}{L_L} \right)\left(\frac{\alpha}{\omega} - t \right) \right\} \right] d(\omega t) \tag{5.8}$$

and

$$I_{L(rms)} = \sqrt{\left[\frac{1}{2\pi} \int_{\alpha}^{\alpha+\gamma} \left[\sin(\omega t - \varphi) - \sin(\alpha - \varphi) \exp\left\{ \left(\frac{R_L}{L_L} \right)\left(\frac{\alpha}{\omega} - t \right) \right\} \right]^2 d(\omega t) \right]} \tag{5.9}$$

Inspection of Equations 5.6 through 5.9 reveals that with changing α, the values of the significant variables such as β, γ, i_L, $I_{L(av)}$, and $I_{L(rms)}$ are changed.

5.3 Walsh Domain Analysis of Half-Wave Controlled Rectifier

To determine the forced response of the circuit given in Figure 5.2a using Walsh domain operational technique, we have two primary tasks:

1. To represent the load voltage waveform $v_o(t)$, which depends on the angles α and φ (which determines the extinction angle β), in terms of Walsh functions
2. To represent the load by a **WOTF**

After completion of these two tasks, we employ Equation 3.41 to determine the desired output.

To perform the first task, we use Equation 5.4 to represent a sine wave by a piecewise constant stepped waveform. Our domain of interest will be 0 to $1/f$ seconds, where f is the frequency of the ideal source voltage. Hence, the scaling constant will be $1/f$. From the load circuit parameters R_L and L_L, we may determine the angle φ, and depending upon the chosen value of α, the extinction angle β may be obtained.

Now, for a particular triggering angle α, the voltage is truncated and the voltage across the load is shown in Figure 5.2b. To represent this wave by a set of 16 block pulse functions, we make use of the area-preserving technique mentioned in Section 4.3.2. First, we compute the angle β, and from the knowledge of α and β, we truncate the stepped waveform (equivalent sine wave) of Figure 5.1b to obtain the stepped load voltage waveform shown in Figure 5.3a. Since we consider the 16 basis functions, the smallest subinterval in our timescale would be $1/16f$ seconds or equivalently 360/16 or 22.5 degrees. However, if, instead of 16, m basis functions are used, the duration of each smallest subinterval would be $1/mf$ seconds, or equivalently 360/m degrees.

If the values of α and β are such that none of the angles is an integral multiple of the smallest subinterval, we shall have difficulty in representing the truncated load voltage in terms of Walsh or block pulse functions in a straightforward manner. This problem is solved with the help of the area-preserving transformation.

As shown in Figure 5.3a, let us suppose that the arbitrary value of α is such that

$$\alpha = \left(\frac{360}{16}\right)[n+x] = 22.5(n+x) \text{ degrees} \tag{5.10}$$

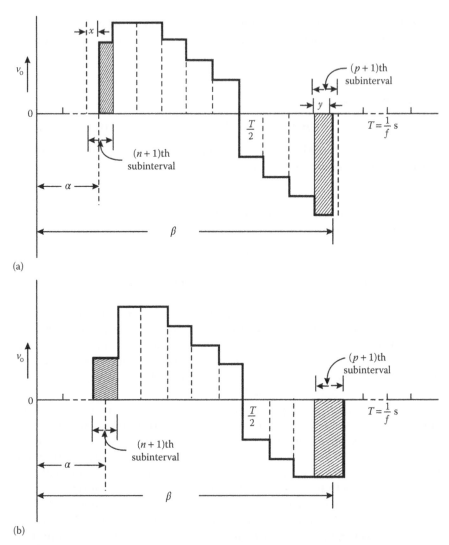

(a)

(b)

FIGURE 5.3
(a) Block pulse domain representation of the normalized load voltage waveform for $m = 16$ and (b) the same voltage waveform after area-preserving transformation. (Reprinted from Deb, A. and Datta, A. K., *Int. J. Electron.*, 79, 861–883, 1995. With permission.)

where:
   n is an integer
   x is a fraction lying between 0 and 1

In the $(n + 1)$th subinterval, the $(n + 1)$th block pulse function covers only $22.5(1 − x)$ degrees instead of the full 22.5-degree subinterval. This is shown by the shaded area in Figure 5.3a.

Knowing the load parameters R_L, L_L, supply voltage, and angular frequency ω, the phase angle φ could be determined. Then, from Equation 5.7, the extinction angle β is obtained.

Similar to α, if β is found to be a nonintegral multiple of the smallest subinterval of 22.5 degrees, we assume that β is given by

$$\beta = 22.5 \ (p+y) \text{ degrees} \tag{5.11}$$

where:

p is an integer

y is a fraction like x lying between 0 and 1

Therefore, the extinction angle terminates in the $(p + 1)$th subinterval as shown in Figure 5.3a. Like α, the fractional overflow of β is also shown by the shaded area of the $(p + 1)$th subinterval.

This truncated piecewise constant stepped waveform is unsuitable, however, for Walsh domain mathematical manipulations. Hence, we apply the area-preserving technique to transform the waveform of Figure 5.3a to that of Figure 5.3b. The transformation is simple in the sense that the shaded areas of Figure 5.3a are made to cover the whole of $(n + 1)$th and $(p + 1)$th subintervals, and their amplitudes are reduced accordingly. It means, the shaded areas in the corresponding subintervals $(n + 1)$ and $(p + 1)$ of Figure 5.3a and 5.3b are equal. This transformation helps us in representing the load voltage waveform in terms of Walsh or block pulse functions. From Figure 5.3b, we may write down an equation similar to Equation 5.4.

From Figure 5.2a, the load transfer function is given by

$$G_4(s) = \frac{Z_L}{R_L + L_L s} \tag{5.12}$$

where:
   s is the Laplace operator

We consider the load voltage as the input variable and the normalized current as the output variable.

To obtain the **WOTF**, we replace the Laplace operator of Equation 5.12 by the operational matrix for differentiation **D**.

Considering m basis functions, we may write

$$\textbf{WOTF4} = Z_L \left[R_L \textbf{I} + L_L \textbf{D} \right]^{-1}$$

where:
   I is a unit matrix of order m
   D is the operational matrix for differentiation of order $(m \times m)$

Since we consider only one cycle of the load voltage, the scaling constant is $1/f$, and from Equation 3.37, **WOTF4**—after scaling—is given by

$$\mathbf{WOTF4} = Z_L[R_L\mathbf{I} + fL_L\mathbf{D}]^{-1} = \left(\frac{Z_L}{fL_L}\right)\left[\left(\frac{R_L}{fL_L}\right)\mathbf{I} + \mathbf{D}\right]^{-1} \quad (5.13)$$

As outlined in Sections 3.3 through 3.5, the normalized current in block pulse domain is given by

$$\mathbf{C'}\Psi_{(m)} = \frac{1}{m}[r_0' \ r_1'...r_{m-2}' \ r_{m-1}']\mathbf{W} \ \{\mathbf{WOTF4}\}\mathbf{W} \ \Psi_{(m)} \quad (5.14)$$

where:

$r_0', r_1',..., r_{m-2}', r_{m-1}'$ are the block pulse coefficients of the input voltage waveform corresponding to Figure 5.3b

W is the Walsh matrix of order $(m \times m)$

For the circuits under investigation, the parameters are

1. $f = 50$ Hz, $\varphi = 45°$, $R_L = 10 \ \Omega$, and $L_L = 0.031831$ (as determined from the values of $R_L, f,$ and φ).
2. $f = 50$ Hz, $\varphi = 60°$, $R_L = 10 \ \Omega$, and $L_L = 0.05513$ (as determined from the values of $R_L, f,$ and φ).
3. $f = 50$ Hz, $\varphi = 75°$, $R_L = 10 \ \Omega$, and $L_L = 0.118795$ (as determined from the values of $R_L, f,$ and φ).

and 16 basis Walsh functions are considered: $m = 16$.

For the above three cases, Equation 5.14 was employed to find out the normalized average and rms currents. Variations of these currents with triggering angle α are shown in Figure 5.4a and 5.4b, and these curves are compared with the exact curves obtained from Equations 5.8 and 5.9 [18]. From Figure 5.4a and 5.4b, it is evident that the Walsh domain solutions are very close to the exact solutions (Program B.23 in Appendix B).

5.3.1 Computational Algorithm

In computing the extinction angle β from Equation 5.7, the well-known Newton–Raphson method [19,20] has been used. For obtaining $\mathbf{C'}\Psi$ for various different values of α, we have developed a fast algorithm that uses a matrix of order (8×16) as the input matrix. Representing Equation 5.4 by the following general form, we have

$$\sin(2\pi ft) \approx [r_0' \ r_1' \ r_2'...r_{14}' \ r_{15}']\Psi_{(16)} \quad (5.15)$$

Now referring to Figure 5.3a and Equation 5.10, if $n = 0$, $\alpha = 22.5x$ degrees, which is equivalent to a delay of $x/16f$ seconds. Hence, for $\alpha = 22.5x$ degrees, the first

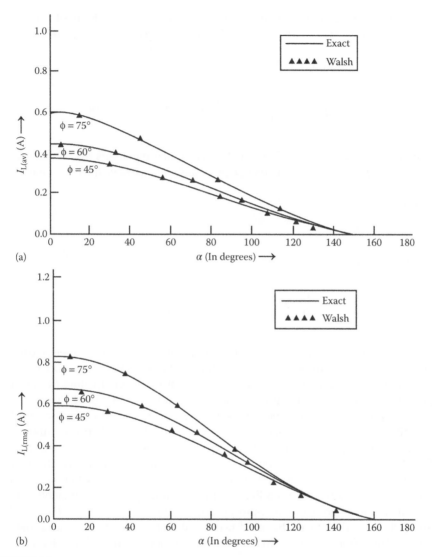

FIGURE 5.4
(a) Variation of the normalized average current with triggering angle α and (b) variation of the normalized rms current with triggering angle α. (Reprinted from Deb, A. and Datta, A. K., *Int. J. Electron.*, 79, 861–883, 1995. With permission.)

coefficient r_0' of Equation 5.15 is transformed to $(1-x)r_0'$, and it fully covers the 22.5 degrees of the first subinterval.

Hence, Equation 5.15 is now converted to

$$\sin(2\pi ft)\left[u\left(t-\frac{x}{16f}\right)\right] \approx \left[(1-x)r_0'\ r_1'\ r_2'...r_{15}'\right]\Psi_{(16)} \tag{5.16}$$

When $n = 1$, $\alpha = 22.5(1 + x)$ degrees that is equivalent to a delay of $1 + x/16f$ seconds; in this case, r_0' of Equation 5.15 becomes 0 and r_1' is transformed to $(1 - x)r_1'$.

In this case, Equation 5.15 takes the form:

$$\sin(2\pi ft)\left[u\left\{t - \left(\frac{1+x}{16f}\right)\right\}\right] \approx \left[0 \ (1 - x)r_1' \ r_2' \ \dots \ r_{15}'\right]\Psi_{(16)} \tag{5.17}$$

Similarly, for $n = 2, 3, 4, \dots, 7$, we shall have similar equations for the truncated sine waveform. Combining all these eight equations, we have the input matrix as

$$\mathbf{R_H'}\Psi_{(16)} =$$

$$\begin{bmatrix} (1-x)r_0' & r_1' & r_2' & & \cdots & & r_{14}' & r_{15}' \\ 0 & (1-x)r_1' & r_2' & & \cdots & & r_{14}' & r_{15}' \\ 0 & 0 & (1-x)r_2' & & \cdots & & r_{14}' & r_{15}' \\ \vdots & \vdots & \vdots & & & & \vdots & \vdots \\ 0 & 0 & 0 & (1-x)r_6' & r_7' & r_8' & \cdots & r_{14}' & r_{15}' \\ 0 & 0 & 0 & \cdots & 0 & (1-x)r_7' & r_8' & \cdots & r_{14}' & r_{15}' \end{bmatrix}_{(8 \times 16)} \Psi_{(16)} \tag{5.18}$$

Inspection of Equation 5.18 reveals that for a particular value of x, eight different values of α are handled by the input matrix. Hence, a variation of x from 0 to 1 by steps of 0.1 (say) will effectively take care of 81 different values of the triggering angle α, and this technique leads to a faster computational algorithm.

In Equation 5.18, the coefficients r_8', r_9',..., r_{14}', r_{15}' are negative, indicating the negative half-cycle of the sine wave represented by block pulse functions. However, our algorithm first determines β to find out the values of p and y in Equation 5.11 from the knowledge of α and, accordingly, determines the actual coefficients r_8'', r_9'',..., r_{14}'', r_{15}''. These coefficients replace the old set of coefficients r_8' to r_{15}' in Equation 5.18 to modify $\mathbf{R_H'}$. In determining these coefficients, some of which may be zero, the area-preserving technique, as shown in Figure 5.3a and 5.3b, is used. We call this modified input matrix the phase control matrix (PCM). Then, from Equation 5.14, the output solution matrix will be given by

$$\mathbf{C'}\Psi_{(16)} = \frac{1}{16}\{\mathbf{PCM}\} \ \mathbf{W} \ \{\mathbf{WOTF4}\} \ \mathbf{W} \ \Psi_{(16)} \tag{5.19}$$

The computational steps involved may briefly be enumerated as follows:

1. Read the input matrix $\mathbf{R_H'}$ for $x = 0$.

   Read the Walsh matrix $\mathbf{W}$.

   Read the Walsh domain operational matrix $\mathbf{D}$.

2. Read R_L, L_L, and f.

3. Compute load impedance Z_L and load phase angle φ.

4. Determine **WOTF4** in the scaled Walsh domain with the help of Equation 5.13.

5. Modify each row of R'_H for a particular chosen value of x.

6. Determine eight different values of the triggering angle α for the value of x chosen in step 5 using Equation 5.10, where $n = 0$ to 7.

7. Compute eight different values of β from Equation 5.7 for eight different values of α, obtained in step 6 by applying Newton–Raphson method.

8. Again modify each row of R'_H for eight different values of β, obtained in step 7, to get the **PCM**.

9. Use Equation 5.19 to evaluate C'.

10. From each row of C', determine the normalized average and rms values of the load currents using the following formulae:

$$I_{L1(av)} = \frac{\text{Sum of all the coefficients in a row}}{16}$$

and

$$I_{L1(rms)} = \sqrt{\frac{\text{Sum of squares of all the coefficients in a row}}{16}}$$

11. Change the value of x and go to step 5 until x reaches its chosen limit, that is, ≤ 1.

12. Stop.

Example 5.1

In a series RL circuit of half-wave controlled rectifier, let $R = 25\ \Omega$ and $L = 150\ \text{mH}$. The supply voltage is given by $e = E_m \sin \omega t$, where $E_m = 400\ \text{V}$ at a frequency of 50 Hz [21].

The conduction angle γ obtained via Newton–Raphson method is $\gamma = 216.166°$ and the extinction angle $\beta = 246.165°$ when $\alpha = 30°$. For $\alpha = 120°$, the conduction angle is $\gamma = 103°$ and the extinction angle is $\beta = 222.991°$.

The results obtained in the Walsh domain are compared with the exact solutions in Table 5.1. The Walsh domain representation of the load current is shown in Figure 5.5 along with the exact solution (Programs B.24 and B.25 in Appendix B).

TABLE 5.1

Comparative Study of Load Currents for Different Triggering Angles

α (In Degrees)	Average Value (A)		rms Value (A)	
	Exact	Walsh Analysis	Exact	Walsh Analysis
For $\alpha = 30°$	3.234	3.194	Too complex to determine	4.612
For $\alpha = 120°$	0.590	0.599	Too complex to determine	1.178

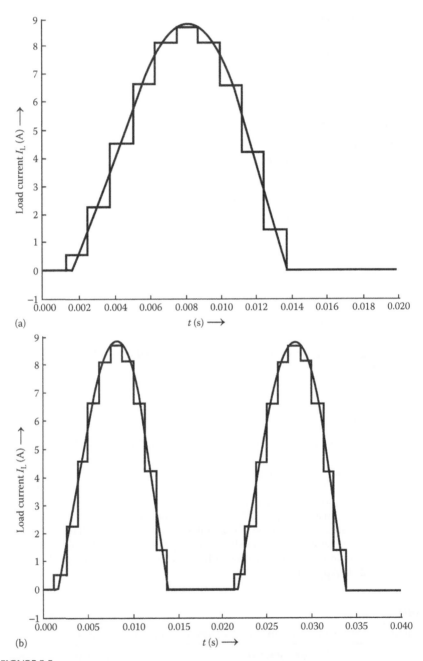

FIGURE 5.5
Walsh domain representation ($m = 16$, $T = 0.02$ seconds) of the load current waveforms I_L along with the exact solutions for (a) one cycle when $\alpha = 30°$, (b) two cycles when $\alpha = 30°$, (c) one cycle when $\alpha = 120°$, and (d) two cycles when $\alpha = 120°$.

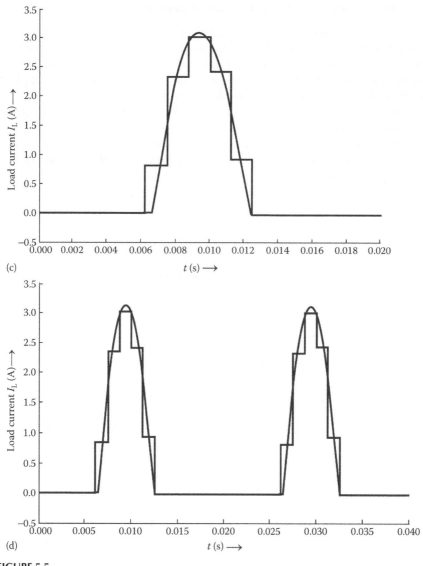

FIGURE 5.5
Continued

5.4 Walsh Domain Analysis of Full-Wave Controlled Rectifier

The analysis of a circuit fed by a single-phase full-wave controlled rectifier proceeds in a line similar to that presented in Section 5.3. Here also, we represent the applied load voltage waveform in terms of Walsh functions and obtain a **WOTF** for the load. Then, the application of Equation 3.41 gives the desired result.

5.4.1 Single-Phase Full-Wave Controlled Rectifier

Figure 5.6 shows two schematic circuit diagrams for single-phase full-wave controlled rectifier feeding RL loads. The circuit of Figure 5.6a represents a bridge rectifier and may be thought to be advantageous with respect to the circuit of Figure 5.6b. This is because the maximum reverse voltage applied to a thyristor in the circuit of Figure 5.6a is only half of that applied to a thyristor in the circuit of Figure 5.6b. However, this advantage must be weighed against extra component costs.

If the transformers and the thyristors of the circuits of Figure 5.6 are regarded as ideal, each of these circuits may be represented by the equivalent circuit of Figure 5.7, where

$$v_{AN} = \sqrt{2}V \sin(\omega t)$$

and

$$v_{BN} = -\sqrt{2}V \sin(\omega t)$$

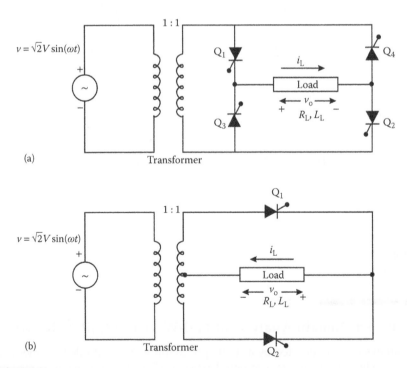

FIGURE 5.6
Single-phase full-wave controlled rectifier circuits feeding an RL load using (a) four thyristors and (b) two thyristors.

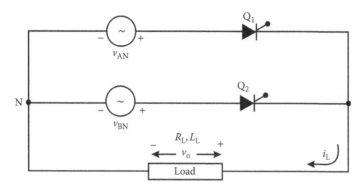

FIGURE 5.7
Equivalent circuit of a single-phase full-wave controlled rectifier feeding an RL load.

5.4.2 Representation of the Load Voltage by Walsh Functions

For a two-pulse rectifier having an equivalent circuit as shown in Figure 5.7, the load voltage $v_{o1}(t)$ could be represented as the product of a sine wave and a piecewise constant function $f_1(t)$ having amplitude +1 or –1.

Considering a typical load voltage waveform $v_{o1}(t)$, we express it as the product of $f_1(t)$ and $\sin(\omega t)$, which is shown in Figure 5.8.

From this figure,

$$f_1(t) = \left[u(t - t_2) - u\left(t - \frac{T}{2} - t_1 \right) \right] - \left[u\left(t - \frac{T}{2} - t_2 \right) - u(t - T) \right] \qquad (5.20)$$

and

$$v_{o1}(t) = [\sin(\omega t)] \, f_1(t) \qquad (5.21)$$

In this figure, the triggering angle α and the extinction angle β are given by

$$\alpha = \omega t_2 \text{ rad}$$

and

$$\beta = \omega \left(\frac{T}{2} + t_1 \right) \text{rad.}$$

It should be noted that β has been computed from the knowledge of R_L, L_L, φ, and α, as indicated by Equation 5.7.

Knowing the particular load voltage waveform $v_{o1}(t)$ from Figure 5.8, we convert it into its equivalent block pulse representation.

First, we use Equation 5.4 to represent the sine wave of Figure 5.8 by a combination of 16 basis block pulse functions. The next step is to mark out the angles α and β on this transformed sine waveform. For typical values of α and β, this has been shown in Figure 5.9a. Now we multiply this waveform

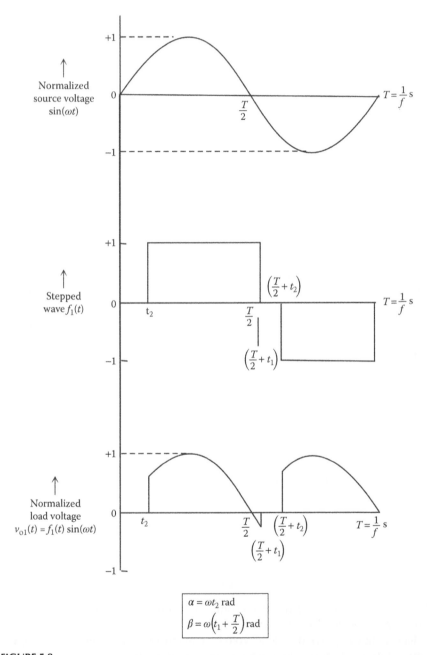

FIGURE 5.8
Representation of the load voltage $v_{o1}(t)$ as a product of $f_1(t)$ and $\sin(\omega t)$. (Reprinted from Deb, A. and Datta, A. K., *Int. J. Electron.*, 79, 861–883, 1995. With permission.)

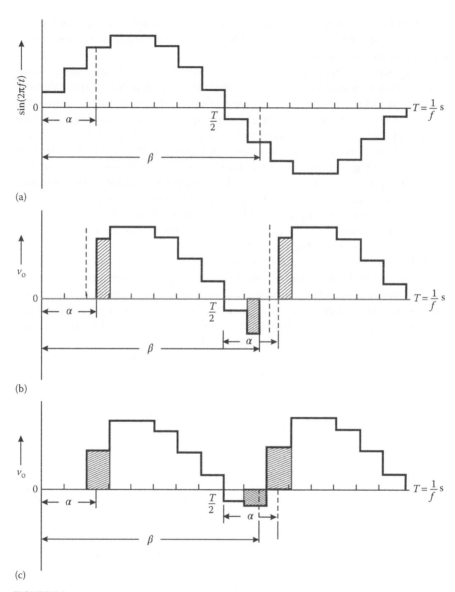

FIGURE 5.9
(a) Block pulse representation of a sine wave with typical angles α and β, (b) block pulse representation of the load voltage waveform $v_{o1}(t)$, and (c) block pulse representation of the load voltage waveform $v_{o1}(t)$ after area-preserving transformation. (Reprinted from Deb, A. and Datta, A. K., *Int. J. Electron.*, 79, 861–883, 1995. With permission.)

by $f_1(t)$ of Figure 5.8 to obtain the load voltage $v_{o1}(t)$ expressed in block pulse domain, as shown in Figure 5.9b. The voltage waveform of Figure 5.9b is equivalent to $v_{o1}(t)$ of Figure 5.8.

As explained in Section 5.3, the waveform of Figure 5.9b may not be suitable for Walsh domain analysis. Its suitability will depend on the values of α and β. If both α and β are integral multiples of the smallest subinterval of $1/16f$ seconds, or equivalently $360/16$ degrees, we can directly represent the waveform in terms of Walsh or block pulse functions and write down a matrix equation similar to Equation 5.4. However, if α and β are arbitrary and not an integral multiple of the smallest subinterval, we have difficulty in direct representation of the load voltage waveform in the Walsh domain. So again, we seek resort to the area-preserving technique to handle the arbitrary values of α and β represented by Equations 5.10 and 5.11.

The application of the area-preserving technique transforms the waveform of Figure 5.9b to that of Figure 5.9c, and now we do not have any hindrance in Walsh/block pulse domain representation of the load voltage waveform.

Figure 5.9b and 5.9c shows the area-preserving transformation of a typical load voltage. For such transformations, the areas under consideration should always be represented by their appropriate signs. Let us suppose that for a particular load voltage waveform, the angles $(\alpha + \pi)$ and β fall in the same subinterval and may have equal or unequal values, which is a situation typical of a two-pulse rectifier. In such case, using Equations 5.10 and 5.11, the angles $(\alpha + \pi)$ and β may be given by

$$\alpha + \pi = \left(\frac{360}{m}\right)\left[n + x + \frac{m}{2}\right] \text{degrees} \qquad (5.22)$$

and

$$\beta = \left(\frac{360}{m}\right)[p + y] \text{degrees} \qquad (5.23)$$

where:
  m is the number of subintervals considered in one cycle of the input
    waveform
  x and y are fractions lying between 0 and 1
  n and p are integers given by

$$0 \le n \le \frac{m}{2}$$

and

$$\frac{m}{2} \le p \le m$$

When $(\alpha + \pi)$ and β fall in the same subinterval, we have, from Equations 5.22 and 5.23,

$$p = n + \frac{m}{2}$$

It is evident that under this situation, the equality of the fractions x and y dictates the equality of the angles $(\alpha + \pi)$ and β. It may also be noted that the angles α and β follow the constraint

$$\beta \le (\alpha + \pi)$$

due to obvious physical reasons.

Hence, if we have $p = n + (m/2)$, in the $(p + 1)$th subinterval, the area-preserving transformation will convert the amplitude r'_p (say) of the block pulse function in the $(p+1)$th subinterval to $[y-(1-x)]\, r'_p$. Since $(p+1) > (m/2)$, the coefficient r'_p of the input waveform will always be negative (conforming to the negative half-cycle of a sine wave). But the transformed amplitude $[y-(1-x)]\, r'_p$ will be negative only when $y > (1 - x)$ and positive when $y < (1 - x)$. If $(x + y) = 1$, the transformed amplitude $[y-(1-x)]\, r'_p$ will be 0.

For other values of α and β, the area-preserving transformation is similar to that for a half-wave rectifier.

Therefore, after the necessary transformation, the load voltage could be represented in terms of Walsh or block pulse functions, and from Figure 5.9c, we can write down an equation similar to Equation 5.4.

5.4.3 Determination of Normalized Average and rms Currents

For the full-wave controlled rectifier, the load transfer function will remain unchanged. So the **WOTF** will again be given by Equation 5.13 and the output $\mathbf{C}' \Psi$, which describes the normalized currents, will be as given by Equation 5.14, where the coefficients $r'_0, r'_1, \ldots, r'_{m-2}, r'_{m-1}$ will have different values.

For the circuit under investigation, the parameters are $R_L = 4.35\ \Omega$, $L_L = 0.02$ H, $f = 50$ Hz, and $\varphi = 55.3°$, and 16 basis Walsh functions were considered: $m = 16$.

To determine the normalized average and rms currents, Equation 5.14 has been used. Results obtained by the Walsh analysis are later compared with the exact solutions for different values of α.

Now, from the shape of the load voltage waveform $v_{o1}(t)$ shown in Figure 5.8, we find that the waveform is not periodic, and from $t = 0$ to $t = t_2$ (or, equivalently, angle α), the load voltage $v_{o1} = 0$. This is due to the initial switching of the thyristors at $\alpha = \omega t_2$ rad. In the next half-cycle (from $T/2$ to T), the voltage waveform is somewhat different from that in the first half-cycle, and it is assumed that in all the forthcoming half-cycles, the waveform of the second half-cycle is repeated periodically. Then, the steady-state waveform would be as shown in Figure 5.10.

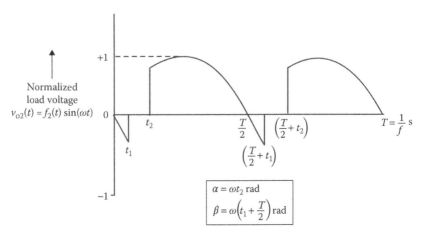

FIGURE 5.10
Steady-state load voltage waveform $v_{o2}(t)$.

If we consider that the waveform of Figure 5.10 is switched across the load at $t = 0$, the load voltage $v_{o2}(t)$ can be expressed as

$$v_{o2}(t) = [\sin(\omega t)]f_2(t) \tag{5.24}$$

where:

$$f_2(t) = -[u(t) - u(t - t_1)] + f_1(t) \tag{5.25}$$

and $f_1(t)$ is given by Equation 5.20. We represent this fact by Figure 5.11, which is similar to Figure 5.8.

As outlined in Section 5.4.2, we transform the waveform of Figure 5.10 using the area-preserving technique and make it suitable for Walsh domain mathematical manipulations. Compared to Figure 5.9b and 5.9c, the only difference that we shall have in this case is that the two half-cycle waves will be identical as shown in Figure 5.10.

With this load voltage waveform, if we carry out the computations with the help of Equation 5.14 considering the same circuit conditions—$R_L = 4.35\ \Omega$, $L_L = 0.02\ H$, $f = 50\ Hz$, $\varphi = 55.3°$, and $m = 16$—the solutions for the normalized average and rms currents can easily be computed by Walsh analysis.

5.4.3.1 Exact Equations for Phase-Controlled Rectifier

Referring to Figure 5.8 and Equation 5.21, the corresponding load voltage waveform is given by

$$v_{o1}(t) = f_1(t) \sin(\omega t)$$

$$= \left[\left\{u(t - t_2) - u\left(t - \frac{T}{2} - t_1\right)\right\} - \left\{u\left(t - \frac{T}{2} - t_2\right) - u(t - T)\right\}\right]\sin(\omega t) \tag{5.26}$$

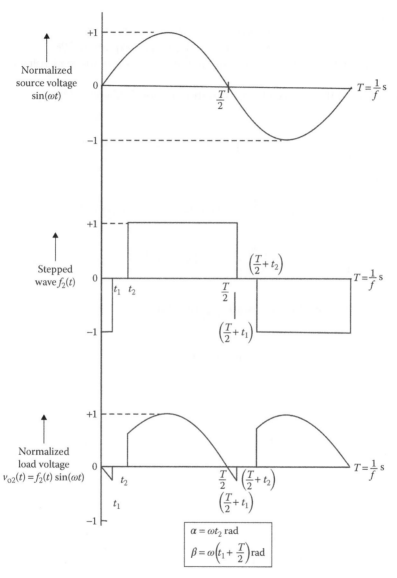

FIGURE 5.11
Representation of the load voltage $v_{o2}(t)$ as a product of $f_2(t)$ and $\sin(\omega t)$. (Reprinted from Deb, A. and Datta, A. K., *Int. J. Electron.*, 79, 861–883, 1995. With permission.)

For a load having a resistance of R_L Ω and an inductance of L_L H, the load current is given by

$$I_{L1}(s) = \frac{E_m V_{o1}(s)}{L_L \left(s + \dfrac{R_L}{L_L} \right)} \tag{5.27}$$

where:

$I_{L1}(s)$ is the Laplace transform of the load current $i_{L1}(t)$
$V_{o1}(s)$ is the Laplace transform of $v_{o1}(t)$ given by Equation 5.26
E_m is a constant factor contributing to the amplitude of the load voltage $v_{o1}(t)$

Evaluation of the Laplace transform in Equation 5.27 yields [22]

$$I_{L1}(s) = \frac{E_m}{L_L\left(s + \dfrac{R_L}{L_L}\right)(s^2 + \omega^2)}\Bigg[\{s\,\sin(\omega t_2) + \omega\cos(\omega t_2)\}\exp(-t_2 s)$$

$$-\left[s\,\sin\left\{\omega\left(t_1 + \frac{T}{2}\right)\right\} + \omega\cos\left\{\omega\left(t_1 + \frac{T}{2}\right)\right\}\right]\exp\left\{-\left(t_1 + \frac{T}{2}\right)s\right\}$$

$$-\left[s\,\sin\left\{\omega\left(t_2 + \frac{T}{2}\right)\right\} + \omega\cos\left\{\omega\left(t_2 + \frac{T}{2}\right)\right\}\right]\exp\left\{-\left(t_2 + \frac{T}{2}\right)s\right\}$$

$$+\{s\,\sin(\omega T) + \omega\cos(\omega T)\}\exp(-Ts)\Bigg] \tag{5.28}$$

The inverse transform of Equation 5.28 yields

$$i_{L1}(t) = \frac{E_m}{Z_L}\Bigg[\left\{-\exp\left[\frac{-R_L(t - t_2)}{L_L}\right]\sin(\omega t_2 - \varphi) + \sin(\omega t - \varphi)\right\}u(t - t_2)$$

$$+\left[\exp\left\{\frac{-R_L\left(t - t_1 - \dfrac{T}{2}\right)}{L_L}\right\}\sin\left(\omega t_1 + \omega\frac{T}{2} - \varphi\right) - \sin(\omega t - \varphi)\right]u\left(t - t_1 - \frac{T}{2}\right)$$

$$+\left[\exp\left\{\frac{-R_L\left(t - t_2 - \dfrac{T}{2}\right)}{L_L}\right\}\sin\left(\omega t_2 + \omega\frac{T}{2} - \varphi\right) - \sin(\omega t - \varphi)\right]u\left(t - t_2 - \frac{T}{2}\right)$$

$$+\left[-\exp\left\{-\frac{R_L(t - T)}{L_L}\right\}\sin(\omega T - \varphi) + \sin(\omega t - \varphi)\right]u(t - T)\Bigg] \tag{5.29}$$

Hence, the normalized average load current is

$$I_{L1(av)} = \left(\frac{Z_L}{E_m}\right)\frac{1}{T}\int_0^T i_{L1}(t)\,dt \tag{5.30}$$

Performing integration within the proper limits and simplifying, we have

$$
I_{L1(av)} = \frac{L_L}{R_L T}\left[-\left\{1 - \exp\left[-\frac{R_L(T-t_2)}{L_L}\right]\right\} \sin(\omega t_2 - \varphi)\right.
$$

$$
+ \left\{1 - \exp\left[-\frac{R_L\left(\frac{T}{2}-t_1\right)}{L_L}\right]\right\} \sin\left(\omega t_1 + \omega\frac{T}{2} - \varphi\right)
$$

(5.31)

$$
+ \left\{1 - \exp\left[-\frac{R_L\left(\frac{T}{2}-t_2\right)}{L_L}\right]\right\} \sin\left(\omega t_2 + \omega\frac{T}{2} - \varphi\right)\right]
$$

$$
+ \frac{1}{\omega T}\left[\cos\varphi + \cos(\omega t_2 - \varphi) - \cos\left(\omega t_1 + \omega\frac{T}{2} - \varphi\right) - \cos\left(\omega t_2 + \omega\frac{T}{2} - \varphi\right)\right]
$$

The above equation is employed to compute the exact solutions for the normalized average currents for load voltage $v_{o1}(t)$. To compute the exact solutions for the normalized rms currents, we can use Equation 5.29 which is the actual solution for the load current $i_{L1}(t)$. However, computations for the exact rms currents are much involved as is seen from the nature of Equation 5.29. But if desired, it could be calculated from Equation 5.29. Thus, only Walsh domain solutions for the normalized rms currents for different values of α are computed.

Equations 5.14 and 5.31 are employed for determining the normalized average currents, both in the Walsh domain and in the exact solution, for different values of α for the load voltage waveform $v_{o1}(t)$ of Figure 5.8 and the load system mentioned earlier. These values are tabulated in Table 5.2. It is noted that the results of Walsh analysis are very close to the exact solutions.

TABLE 5.2

Comparison of Walsh Domain Solution of the Normalized Average Current for Different Values of α with the Exact Solution for the Load Voltage Waveform $v_{o1}(t)$ as Shown in Figure 5.8 ($R_L = 4.35\ \Omega$, $L_L = 0.02\ H$, $f = 50\ Hz$, $m = 16$)

	Normalized Average Load Current $I_{L1(av)}$		
α (In Degrees)	Walsh Analysis	Exact	Error (%)
0.0	0.8826	0.8840	0.158
7.5	0.8633	0.8775	1.618
15.0	0.8439	0.8580	1.643
22.5	0.8246	0.8263	0.206
30.0	0.7712	0.7831	1.520

(Continued)

TABLE 5.2

Continued

α (In Degrees)	Normalized Average Load Current $I_{L1(av)}$		
	Walsh Analysis	Exact	Error (%)
37.5	0.7179	0.7295	1.590
45.0	0.6645	0.6667	0.330
52.5	0.5881	0.5961	1.342
60.0	0.5282	0.5360	1.455
67.5	0.4786	0.4827	0.849
75.0	0.4229	0.4276	1.100
82.5	0.3679	0.3721	1.129
90.0	0.3136	0.3173	1.166
97.5	0.2638	0.2644	0.227
105.0	0.2149	0.2144	−0.233
112.5	0.1647	0.1684	2.197
120.0	0.1278	0.1272	−0.472

Table 5.3 tabulates the normalized rms currents for different values of α for the same system using only Equation 5.14 pertaining to Walsh analysis. This is because of the reasons already mentioned above. However, from the results of Table 5.2, it can be assumed that the results of Table 5.3 are acceptable (Program B.26 in Appendix B).

TABLE 5.3

Variation of Walsh Domain Solution of Normalized rms Current for Different Values of α for the Load Voltage Waveform $v_{o1}(t)$ as Shown in Figure 5.8 ($R_L = 4.35\ \Omega$, $L_L = 0.02\ H$, $f = 50\ Hz$, $m = 16$)

α (In Degrees)	Normalized rms Load Current via Walsh Analysis $I_{L1(rms)}$
0.0	0.9544
7.5	0.9373
15.0	0.9205
22.5	0.9039
30.0	0.8537
37.5	0.8057
45.0	0.7601
52.5	0.6861
60.0	0.6322
67.5	0.5903
75.0	0.5318
82.5	0.4763
90.0	0.4245

(Continued)

TABLE 5.3

Continued

α (In Degrees)	Normalized rms Load Current via Walsh Analysis $I_{L1(rms)}$
97.5	0.3646
105.0	0.3074
112.5	0.2526
120.0	0.2015

Referring to Figure 5.11 and Equation 5.24, the load voltage waveform may be expressed by

$$v_{o2}(t) = f_2(t) \sin(\omega t)$$

$$= \left[-\{u(t) - u(t - t_1)\} + \left\{ u(t - t_2) - u\left(t - \frac{T}{2} - t_1 \right) \right\} \right.$$

$$\left. - \left\{ u\left(t - \frac{T}{2} - t_2 \right) - u(t - T) \right\} \right] \sin(\omega t) \tag{5.32}$$

Comparing the above equation with Equation 5.26, we note that only the first term $[-\{u(t) - u(t - t_1)\}]$ of Equation 5.32 is absent in Equation 5.26. Hence, in the solution of the load current $i_{L2}(t)$, only the effect of the additional first two terms involving $u(t)$ and $u(t - t_1)$ will have to be considered in Equation 5.29.

Thus, the solution for load current is given by

$$i_{L2}(t) = \frac{E_m}{Z_L} \left[\left\{ -\exp\left(-\frac{R_L t}{L_L} \right) \sin\varphi - \sin(\omega t - \varphi) \right\} u(t) \right.$$

$$+ \left[-\exp\left\{ -\frac{R_L(t - t_1)}{L_L} \right\} \sin(\omega t_1 - \varphi) + \sin(\omega t - \varphi) \right] u(t - t_1)$$

$$+ \left[-\exp\left\{ -\frac{R_L(t - t_2)}{L_L} \right\} \sin(\omega t_2 - \varphi) + \sin(\omega t - \varphi) \right] u(t - t_2)$$

$$+ \left[\exp\left\{ -\frac{R_L\left(t - t_1 - \frac{T}{2} \right)}{L_L} \right\} \sin\left(\omega t_1 + \omega\frac{T}{2} - \varphi \right) - \sin(\omega t - \varphi) \right] u\left(t - t_1 - \frac{T}{2} \right) \tag{5.33}$$

$$+ \left[\exp\left\{ -\frac{R_L\left(t - t_2 - \frac{T}{2} \right)}{L_L} \right\} \sin\left(\omega t_2 + \omega\frac{T}{2} - \varphi \right) - \sin(\omega t - \varphi) \right] u\left(t - t_2 - \frac{T}{2} \right)$$

$$+ \left[-\exp\left\{ -\frac{R_L(t - T)}{L_L} \right\} \sin(\omega T - \varphi) + \sin(\omega t - \varphi) \right] u(t - T) \right]$$

where all other considerations for deriving Equation 5.29 are kept similar.
Hence, the normalized average load current is

$$I_{L2(av)} = \left(\frac{Z_L}{E_m}\right)\frac{1}{T}\int_0^T i_{L2}(t)\,dt \tag{5.34}$$

Performing integration within the proper limits and simplifying, we have

$$
\begin{aligned}
I_{L2(av)} = \frac{L_L}{R_L T}\Bigg[&-\left\{1-\exp\left(-\frac{R_L T}{L_L}\right)\right\}\sin\varphi - \left[1-\exp\left\{-\frac{R_L(T-t_1)}{L_L}\right\}\right]\sin(\omega t_1 - \varphi) \\
&-\left[1-\exp\left\{\frac{-R_L(T-t_2)}{L_L}\right\}\right]\sin(\omega t_2 - \varphi) \\
&+\left[1-\exp\left\{-\frac{R_L\left(\frac{T}{2}-t_1\right)}{L_L}\right\}\right]\sin\left(\omega t_1 + \omega\frac{T}{2}-\varphi\right) \\
&+\left[1-\exp\left\{\frac{-R_L\left(\frac{T}{2}-t_2\right)}{L_L}\right\}\right]\sin\left(\omega t_2 + \omega\frac{T}{2}-\varphi\right)\Bigg] \\
&+\frac{1}{\omega T}\left[\cos(\omega t_1-\varphi)+\cos(\omega t_2-\varphi)-\cos\left(\omega t_1+\omega\frac{T}{2}-\varphi\right)-\cos\left(\omega t_2+\omega\frac{T}{2}-\varphi\right)\right]
\end{aligned}
\tag{5.35}
$$

The above equation is used to compute the exact solutions for the normalized average currents for the load voltage $v_{o2}(t)$. Here also, we did not attempt
to compute the exact values of the normalized rms currents due to many
involved mathematical operations. But, if desired, it could be calculated from
Equation 5.33. Thus, only Walsh domain solutions for the normalized rms
currents for different values of α are computed.

Equation 5.35 is used to find the values of the normalized average currents
for different triggering angles that are compared with Walsh domain solutions and presented in Table 5.4.

Equations 5.14 and 5.35 are employed for determining the normalized
average currents, both in the Walsh domain and in the exact solution, for different values of α for the load voltage waveform $v_{o2}(t)$ of Figure 5.11 and the
load system mentioned earlier. These values are tabulated in Table 5.4. It is
noted that the results of Walsh analysis are very close to the exact solutions.

Table 5.5 tabulates the normalized rms currents for different values of α
for the same system using Equation 5.14 in the Walsh domain only. This
is because computation of the rms currents is very complex and tedious.
However, from the results of Table 5.4, it can be assumed that the results of
Table 5.5 are reliable (Program B.27 in Appendix B).

TABLE 5.4

Comparison of Walsh Domain Solution of Normalized Average Current for Different Values of α with the Exact Solution for the Load Voltage Waveform $v_{o2}(t)$ as Shown in Figure 5.10 ($R_L = 4.35\ \Omega$, $L_L = 0.02$ H, $f = 50$ Hz, $m = 16$)

	Normalized Average Load Current $I_{L2(av)}$		
α (In Degrees)	Walsh Analysis	Exact	Error (%)
0.0	0.8826	0.8840	0.158
7.5	0.8563	0.8751	2.148
15.0	0.8299	0.8486	2.204
22.5	0.8036	0.8053	0.211
30.0	0.7304	0.7463	2.130
37.5	0.6572	0.6727	2.304
45.0	0.5841	0.5863	0.375
52.5	0.4782	0.4889	2.188
60.0	0.4095	0.4202	2.546
67.5	0.3640	0.3710	1.887
75.0	0.3133	0.3207	2.307
82.5	0.2641	0.2707	2.438
90.0	0.2166	0.2221	2.476
97.5	0.1745	0.1762	0.965
105.0	0.1345	0.1340	−0.373
112.5	0.0910	0.0965	5.699
120.0	0.0618	0.0644	4.037

TABLE 5.5

Variation of Walsh Domain Solution of Normalized rms Current for Different Values of α for the Load Voltage Waveform $v_{o2}(t)$ as Shown in Figure 5.10 ($R_L = 4.35\ \Omega$, $L_L = 0.02$ H, $f = 50$ Hz, $m = 16$)

α (In Degrees)	Normalized rms Load Current via Walsh Analysis $I_{L2(rms)}$
0.0	0.9544
7.5	0.9331
15.0	0.9122
22.5	0.8918
30.0	0.8282
37.5	0.7683
45.0	0.7130
52.5	0.6199
60.0	0.5658
67.5	0.5336
75.0	0.4831

(Continued)

TABLE 5.5

Continued

α (In Degrees)	Normalized rms Load Current via Walsh Analysis $I_{L2(rms)}$
82.5	0.4368
90.0	0.3953
97.5	0.3425
105.0	0.2933
112.5	0.2455
120.0	0.1984

5.4.4 Computational Algorithm

The algorithm developed for the computation of the normalized average and rms currents in the Walsh domain for a two-pulse rectifier is somewhat similar to that described in Section 5.3.1. The extinction angle β was computed from Equation 5.7 by using Newton–Raphson method.

To obtain the **PCM** considering 16 basis functions, we start with an input matrix $\mathbf{R}'_F \Psi$ of the order (8 × 16), each row of which is

$$[r'_0 \; r'_1 \; \dots \; r'_{14} \; r'_{15}] \, \Psi_{(16)}$$

as per Equation 5.15. Then, each row of $\mathbf{R}'_F \Psi_{(16)}$ represents a sine wave generated by block pulse functions as shown in Figure 5.1b. Now depending on the chosen value of α (i.e., time t_2 of Figures 5.8 and 5.11), the extinction angle β is computed from Equation 5.7. From β, the time t_1 of Figure 5.11 is known. From the knowledge of t_1 and t_2, the waveform of Figure 5.9a is converted to that of Figure 5.9b. This is again transformed to the waveform of Figure 5.9c using the area-preserving technique.

If we choose the load voltage waveform $v_{o2}(t)$ similar to that of Figure 5.10, the voltage waveform shown in Figure 5.9b will have identical half-cycles in accordance with the wave shape shown in Figure 5.10. Then, following the rules outlined in Section 5.4.2, the input load voltage could be represented properly by Walsh/block pulse functions.

For the input matrix $\mathbf{R}'_F \Psi$, we choose different values of α for its different rows. This is done with the help of Equation 5.10, and we modify this equation as

$$\alpha = 22.5 \, [(i-1)+x] \text{ degrees} \tag{5.36}$$

where:
i is the ith row of the input matrix $\mathbf{R}'_F \Psi$

Thus, we see that for a particular chosen value of x, eight different values of α are assigned to eight rows of the input matrix depending on the value of i. Therefore, the value of α for the fourth row (say) will be

$$\alpha = 22.5(3 + x) \text{ degrees}$$

This helps to develop a fast algorithm, and each coefficient of each row of $\mathbf{R'_F}$ is modified depending on the values of α and β. Finally, we obtain the modified input **PCM**. Then, Equation 5.19 is used to obtain $\mathbf{C' \Psi}$ and consequently the average and rms values of the normalized current.

The computational steps involved may briefly be enumerated as follows:

1. Read the input matrix $\mathbf{R'_F}$ for $x = 0$.

   Read the Walsh matrix **W**.

   Read the Walsh domain operational matrix **D**.

2. Read R_L, L_L, and f.

3. Compute load impedance Z_L and load phase angle φ.

4. Determine **WOTF4** in the scaled Walsh domain with the help of Equation 5.13.

5. Choose a particular value of x and determine the eight different values of the triggering angle α for the chosen value of x as per Equation 5.10.

6. Compute eight different values of β from Equation 5.7 for eight different values of α, obtained in step 5, by applying Newton–Raphson method.

7. Check whether $\beta > (\alpha + \pi)$. If so, make $\beta = (\alpha + \pi)$ to follow the constraint $\beta \le (\alpha + \pi)$ as mentioned in Section 5.4.2.

8. Modify each row of $\mathbf{R'_F}$ for eight different values of β, obtained in steps 6 and 7, to get the **PCM**.

9. Use Equation 5.19 to evaluate **C'**.

10. From each row of **C'**, determine the normalized average and rms values of load current using the following formulae:

$$I_{L2(av)} = \frac{\text{Sum of all the coefficients in a row}}{16}$$

and

$$I_{L2(rms)} = \sqrt{\frac{\text{Sum of squares of all the coefficients in a row}}{16}}$$

11. Change the value of x and go to step 5 until x reaches its chosen limit, that is, ≤ 1.

12. Stop.

5.5 Conclusion

This chapter presents the Wash analysis of phase-controlled rectifier circuits, and the results obtained by such analysis are presented in Tables 5.2 through 5.5.

It is seen from Table 5.2 that the maximum percentage error for Walsh domain solution of the normalized average current is only 2.197% for $\alpha = 112.5°$ and the error is a minimum of 0.158% for $\alpha = 0°$.

Table 5.3 presents the Walsh domain solution of the normalized rms currents for different triggering angles α, and from inspection of the error column of Table 5.2, it may be conjectured (because the exact solution could not be computed due to complexity) that the solutions presented in Table 5.3 will also be reasonably close to the exact solutions.

From Table 5.4, it is observed that the maximum percentage error for the Walsh domain solution of the normalized average current is only 5.699%, and for triggering angles 0° to 105°, the percentage error is within 2.546%. Here also, the minimum percentage error of 0.158% occurs at $\alpha = 0°$.

The exact solutions of the normalized rms currents for different triggering angles α are too involved to determine. Hence, Table 5.5 presents only the solutions of the normalized rms currents obtained via Walsh analysis. But, as earlier, we can say that deviations of these results from the exact solutions will not be appreciable.

From Tables 5.2 through 5.5, it is noted that the reliable results were obtained by Walsh analysis for a range of values of α from 0° to 120°. Keeping the phase control range of practical controlled rectifier circuits in mind, we can say that the Walsh domain approach is an effective method for the analysis of such circuits.

In this method, the solutions obtained include switching transients that are difficult to realize by conventional methods such as Laplace transforms. To obtain the variables at steady state by this approach, it is necessary to choose a higher value of m so that four to eight cycles of the load voltage wave of Figure 5.9 could be taken into account. Thus, if m is chosen to be 64, 128, or 256, the switching transients in the solution will become insignificant with respect to the steady-state solution. In such case, we may have to handle matrices of the order of (8×128), (128×128) or (8×256), (256×256), and so on. Considering the number-crunching capability of modern computers, this surely does not pose any difficulty.

Still another point to note is the computation of β from Equation 5.7 and its subsequent use for solving the normalized currents both in traditional and in Walsh analyses.

Very small error may have crept in while computing the values of β numerically. But, since the same set of values of β was used for both the analyses, namely, Walsh domain solution and the exact solution, it is evident that error incurred in computation of β will affect both the results in the same way. Thus, Tables 5.2 and 5.4, comparing the results obtained via Walsh and exact analyses, may be regarded as *unaffected* by the errors in β.

References

1. Möltgen, G., *Converter Engineering: An Introduction to Operation and Theory*, John Wiley & Sons, New York, 1984.
2. Rashid, M. H., *Power Electronics: Circuits, Devices, and Applications*, Prentice Hall, Englewood Cliffs, NJ, 1993.
3. Mohan, N., Undeland, T. M. and Robbins, W. P., *Power Electronics: Converters, Applications, and Design* (3rd Ed.), John Wiley & Sons, New York, 2003.
4. Bauer, J., *Single-Phase Pulse-Width Modulated Rectifier*, Acta Polytech., Vol. **48**, No. 3, pp. 84–87, 2008.
5. Abou-Faddan, H. I., Analysis of slip-ring induction motors controlled by silicon-controlled rectifiers under unbalanced stator conditions, *Electric Power Syst. Res.*, Vol. **11**, No. 2, pp. 139–146, 1986.
6. Rodríguez, J. R., Pontt, J., Silva, C., Wiechmann, E. P., Hammond, P. W., Santucci, F. W., Álvarez, R., Musalem, R., Kouro, S. and Lezana, P., Large current rectifiers: State of the art and future trends, *IEEE Trans. Indust. Electron.*, Vol. **52**, No. 3, pp. 738–746, 2005.
7. Patil, P. M. and Kurkute, S. L., Microprocessor based power factor improvement for single phase controlled rectifiers using symmetrical angle control, Proceedings of National Conference on Power Electronics and Power Systems (EtEE'05), St. Joseph's College of Engineering, Chennai, India, January 28–29, 2005, pp. 192–195.
8. Wall, R. W. and Hess, H. L., Design and microcontroller implementation of a three phase SCR power converter, *J. Circuits Syst. Comput.*, Vol. **6**, No. 6, pp. 619–634, 1996.
9. Balamurugan, R. and Gurusamy, G., Harmonic optimization by single phase improved power quality AC-DC power factor corrected converters, *Int. J. Comput. Appl.*, Vol. **1**, No. 5, pp. 33–40, 2010.
10. Carbone, R., A passive power factor correction technique for single-phase thyristor-based controlled rectifiers, *Int. J. Circuits Syst. Signal Process.*, Vol. **2**, No. 3, pp. 169–179, 2008.
11. McCarty, M., Taufik, T., Pratama, A. and Anwari, M., Harmonic analysis of input current of single-phase controlled bridge rectifier, 2009 IEEE Symposium on Industrial Electronics and Applications (ISIEA 2009), Kuala Lumpur, October 4–6, 2009.
12. Daut, I., Ali, R. and Taib, S., Design of a single-phase rectifier with improved power factor and low THD using boost converter technique, *Am. J. Appl. Sci.*, Vol. **3**, No. 7, pp. 1902–1904, 2006.
13. Deb, A. and Datta, A. K., On Walsh/block pulse domain analysis of power electronic circuits: Part 1—Continuously phase-controlled rectifier. *Int. J. Electron.*, Vol. **79**, No. 6, pp. 861–883, 1995.
14. Deb, A. and Datta, A. K., On analytical techniques of power electronic circuit analysis. *IETE Tech. Rev.*, Vol. **7**, No. 1, pp. 25–32, 1990.
15. Schaeffer, J., *Rectifier Circuits: Theory and Design*, John Wiley & Sons, New York, 1965.
16. Davis, R. M., *Power Diode and Thyristor Circuits*, Cambridge University Press, Cambridge, 1971.

17. Pelly, B. R., *Thyristor Phase Controlled Converters and Cycloconverters*, Wiley-Interscience, New York, 1971.
18. Dewan, S. B. and Straughen, A., *Power Semiconductor Circuits*, John Wiley & Sons, New York, 1975.
19. Krishnamurthy, E. V. and Sen, S. K., *Numerical Algorithms*, Affiliated East-West Press, New Delhi, 1986.
20. Demidovich, B. P. and Maron, I. A., *Computational Mathematics*, MIR Publishers, Moscow, 1987.
21. Shepherd W. and Zhang L., *Power Converter Circuits*, Taylor & Francis, New York, 2010.
22. Nixon, F. E., *Handbook of Laplace Transformation* (2nd Ed.), Prentice Hall, Englewood Cliffs, NJ, 1965.

6

Analysis of Inverter Circuits

Inverters convert DC to AC in order to feed power to an AC load from a DC system at some desired voltage and frequency [1–4]. Since the output from an inverter is AC, it can be used in AC drives such as three-phase induction motors or synchronous motors [5]. Also, pulse-width modulated power amplifiers consist of thyristor-bridge inverters [6].

Some of the applications of an inverter are as follows:

1. Inverter-fed AC motors in industries
2. Stand-by power supplies
3. Uninterruptible power supply (UPS) for important equipments, for example, computers
4. Power supplies at frequencies other than conventional supply frequencies for special applications such as aircraft, railways, and communication lines

In the application field of power electronic (PE) circuits, the inverter has much importance, warranting an extensive research such as current control techniques for three-phase voltage source pulse-width modulation (PWM) converter [7] or analysis of current controllers for voltage source inverter [8].

The general block diagram of an inverter system is shown in Figure 6.1. In this figure, the DC source supplies power to the input side of the inverter. The DC filter on the input side is employed to suppress the reverse transmittal of harmonics back into the DC source. At the output of the inverter, an AC voltage is available that is, in general, not sinusoidal. If the load system needs a sinusoidal voltage, it is necessary to connect a suitable filter between the inverter and the AC load as shown in Figure 6.1.

For control of amplitude and frequency of the inverter output voltage, the firing and regulating circuits are necessary. There are many methods available for controlling the frequency and the amplitude of the voltage [9].

When a single-phase AC output is required from a DC source, the commonly employed circuit is the full-bridge inverter circuit. The circuit diagram of the same is shown in Figure 6.2a. The freewheeling diodes are connected in antiparallel with the thyristors. If the load circuit has some inductance, the load current i_L will not reverse at the same instants as does the load

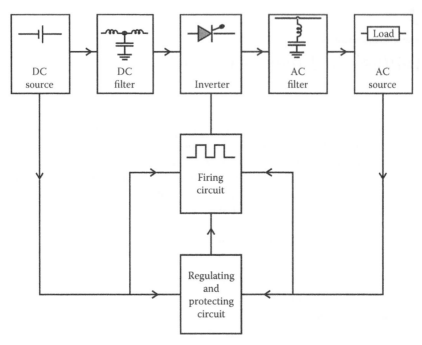

FIGURE 6.1
General block diagram representation of an inverter system.

voltage v_o. The freewheeling diodes permit the load current to flow without reversal at the instants of load voltage reversal. The necessary gating signals for the thyristors and the resultant output voltage waveform are shown in Figure 6.2b.

Pulse-fed PE circuits can be analyzed using Walsh functions [10,11]. This orthogonal function set has special advantage of handling theoretically the modulation of pulse width of a system [12]. PE circuits, especially continuously pulse-width modulated inverters, are studied and discussed with Walsh function as well as block pulse function (BPF) as analysis tools [12].

In what follows, we study the nature of load current in an inverter-fed RL load system when the load voltage is controlled. The analysis is carried out in the Walsh domain, and the results are compared with the exact solutions.

6.1 Voltage Control of a Single-Phase Inverter

In many inverter applications, the control of output voltage [6,12–14] is necessary. The methods employed for such voltage control can be classified into the following three broad categories:

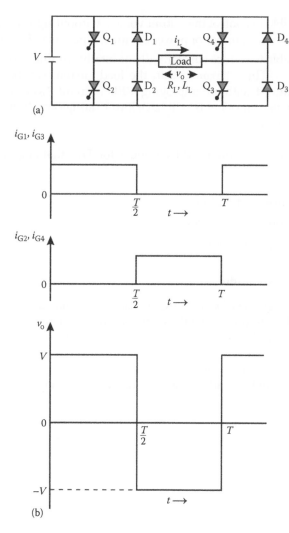

FIGURE 6.2
(a) Full-bridge inverter circuit; (b) gate signals and load voltage waveform of a full-bridge inverter circuit.

1. Control of input DC voltage of the inverter
2. Control of output AC voltage of the inverter
3. Control of voltage within the inverter by the variation of the ratio between the DC input voltage and the AC output voltage of the inverter itself

The third method of control is known as time ratio control (TRC). Inverters employing such control are called pulse-width modulated inverters. The advantages of this method are as follows:

1. It is possible to control the output voltage without significantly adding to the total number of power circuit components of the inverter.

2. It is possible to reduce substantially or totally eliminate lower order harmonics. Thus, harmonics in the load current can be substantially reduced without employing a filter circuit. Even if a filter is required, its components will have lower ratings and therefore will be of lesser cost.

The three most commonly used techniques for TRC that employ PWM are listed as follows:

1. Single-pulse modulation
2. Multiple-pulse modulation
3. Sinusoidal-pulse modulation

6.1.1 Single-Pulse Modulation

The output voltage waveform for a single-pulse modulated inverter is shown in Figure 6.3. It is assumed that the start of each voltage pulse is delayed and the end of each pulse is advanced by an equal time interval of t_1 seconds, resulting in a variation of the pulse width $[(T/2) - 2t_1]$ over the range $0 \leq [(T/2) - 2t_1] \leq (T/2)$ seconds.

The waveform of the load voltage $v_o(t)$ of Figure 6.3 may be given by

$$v_o(t) = \sum_{n=1,3,5,\ldots}^{\infty} \frac{4V}{n\pi} \sin\left[n\pi \frac{\left(\frac{T}{2} - 2t_1 \right)}{T} \right] \sin\left(\frac{2\pi nt}{T} \right) \qquad (6.1)$$

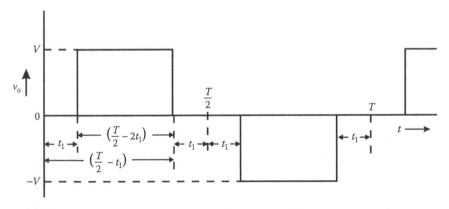

FIGURE 6.3

Single-pulse modulation of the inverter output voltage. (Reprinted from Deb, A. and Datta, A. K., *Int. J. Electron.*, 79, 885–895, 1995. With permission.)

When $t_1 = 0$, we get the alternating square waveform of Figure 6.2b, and when $t_1 = T/4$, the inverter output voltage is 0. As t_1 is increased, the harmonic content of the output voltage waveform increases.

Now we consider that the voltage $v_o(t)$, as shown in Figure 6.3, is impressed across an RL load. Restricting our interest to one cycle only, the load voltage $v_o(t)$ is given by

$$v_o(t) = V\left[u(t-t_1) - u\left(t+t_1-\frac{T}{2}\right) - u\left(t-t_1-\frac{T}{2}\right) + u(t+t_1-T)\right] \quad (6.2)$$

If R_L and L_L are the load parameters, the Laplace transform of the load current is given by

$$I_L(s) = \frac{V}{s(R_L + L_L s)}$$

$$\left[\exp(-t_1 s) - \exp\left\{-\left(\frac{T}{2}-t_1\right)s\right\} - \exp\left\{-\left(\frac{T}{2}+t_1\right)s\right\} + \exp\{-(T-t_1)s\}\right] \quad (6.3)$$

Taking the inverse transform and simplifying, the normalized load current is given by

$$i_L(t) = \left[1-\exp\left(-\frac{R_L t}{L_L}\right)\right]\left[u(t-t_1) - u\left(t-\frac{T}{2}+t_1\right) - u\left(t-\frac{T}{2}-t_1\right) + u(t-T+t_1)\right] \quad (6.4)$$

Hence, the normalized average current is obtained as

$$I_{L(av)} = \frac{2}{T}\int_0^{\frac{T}{2}} i_L(t)dt \quad (6.5)$$

Substituting $i_L(t)$ from Equation 6.4 in Equation 6.5 and performing integrations within proper limits, we have

$$I_{L(av)} = \frac{2}{T}\left(\left(\frac{T}{2}-2t_1\right) + \left(\frac{L_L}{R_L}\right)\left\{\exp\left[-\frac{R_L\left(\frac{T}{2}-t_1\right)}{L_L}\right] - \exp\left(-\frac{R_L t_1}{L_L}\right)\right\}\right) \quad (6.6)$$

Knowing the values of R_L, L_L, t_1, and T, the normalized average load current $I_{L(av)}$ could easily be computed for the cycle(s) under consideration.

6.1.1.1 Walsh Function Representation of Single-Pulse Modulation

Referring to Figure 6.3, we see that it is not difficult to represent the load voltage waveform by Walsh functions. For such representation, we follow the procedure outlined in Section 4.3.2 and employ the area-preserving technique once again. Figure 6.4 shows area-preserving transformation

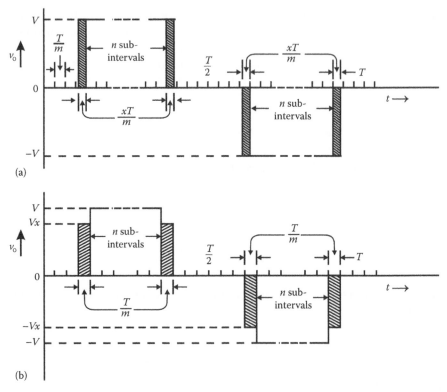

FIGURE 6.4

Single-pulse modulated load voltage waveform $v_o(t)$ (a) before and (b) after area-preserving transformation. (Reprinted from Deb, A. and Datta, A. K., *Int. J. Electron.*, 79, 885–895, 1995. With permission.)

of a typical load voltage v_o. We consider m basis block pulse functions, and hence we need to divide the total interval of interest T seconds into m number of subintervals of equal (T/m)-second duration. For the waveform shown in Figure 6.4a, we see that each half of the wave covers n number of subintervals and a fraction of x ($0 \leq x \leq 1$) of the $[(m/4) - (n/2)]$th subinterval and $[(m/4) + (n/2) + 1]$th subinterval. The fractional overflow on both sides is shown by the shaded portions of the waveform that have a duration of $[x(T/m)]$ seconds each. It may be noted that when the pulse is symmetrically placed in the interval $0 \leq t \leq T/2$, the number n, like m, is always even.

The area-preserving transformation converts the fractional overflows on both sides of the pulse to pulse slices of height xV and width (T/m) seconds, as shown in Figure 6.4b. Now this load voltage waveform can be represented in terms of block pulse functions simply by inspection and is given by

$$v_0(t) \approx \mathbf{R1'}\ \Psi_{(m)} = \left[\underbrace{0\ 0\ 0\ \cdots\ 0\ 0\ 0}_{\left(\frac{m}{4} - \frac{n}{2} - 1\right)\ \mathrm{terms}}\ xV\ \underbrace{V\ V\ \cdots\ V\ V}_{n\ \mathrm{terms}}\ xV \right.$$

$$\underbrace{0\ \cdots\ 0\ 0\ 0\ 0\ \cdots\ 0}_{\left(\frac{m}{2} - n - 2\right)\ \mathrm{terms}}\ -xV\ \underbrace{-V\ -V\ \cdots\ -V\ -V}_{n\ \mathrm{terms}} \quad (6.7)$$

$$\left. -xV\ \underbrace{0\ 0\ 0\ \cdots\ 0\ 0\ 0}_{\left(\frac{m}{4} - \frac{n}{2} - 1\right)\ \mathrm{terms}} \right]\Psi_{(m)}$$

where:
the total number of columns (or terms) in **R1'** is m

Equation 6.7 is the generalized expression for one cycle of the inverter output voltage.

Let for a waveform $\mathbf{R2'}\ \Psi_{(m)}$ under consideration $m = 16$ and $n = 4$. Then Equation 6.7 is transformed to

$$\mathbf{R2'}\ \Psi_{(16)} = \begin{bmatrix} 0 & xV & V & V & V & V & xV & 0 & 0 & -xV & -V & -V & -V & -V & -xV & 0 \end{bmatrix}\Psi_{(16)} \quad (6.8)$$

This is represented in Figure 6.5.

Hence, by varying the number n and the magnitude of the fraction x in Equation 6.7, we can change the width of a pulse continuously and thus

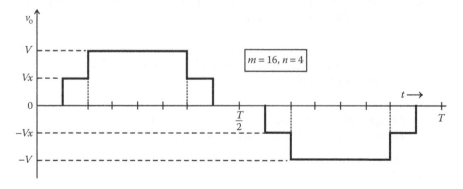

FIGURE 6.5
Inverter load voltage waveform after area-preserving transformation for $n = 4$ and $m = 16$. (Reprinted from Deb, A. and Datta, A. K., *Int. J. Electron.*, 79, 885–895, 1995. With permission.)

achieve continuous single-pulse modulation. When $n = m/2$, we get the waveform of Figure 6.2b, and comparing with Figure 6.3, we find that under the condition $t_1 = 0$ and when $n = 0$, $t_1 = T/4$, the output voltage is 0.

To accommodate all possible variations of n within the limit of $0 < n < m/2$, we can modify Equation 6.7 to the following general matrix equation form:

$$\mathbf{R'}\,\Psi_{(m)} = \left.\underset{\left(\frac{m}{4}\right)\text{rows}}{\left\{\begin{bmatrix} xV & V & V & \cdots & V & V & xV & -xV & -V & -V & \cdots & -V & -V & -xV \\ 0 & xV & V & \cdots & V & xV & 0 & 0 & -xV & -V & \cdots & -V & -xV & 0 \\ 0 & 0 & xV & \cdots & xV & 0 & 0 & 0 & 0 & -xV & \cdots & -xV & 0 & 0 \\ \vdots & \vdots & \vdots & & \vdots & \vdots & \vdots & \vdots & \vdots & \vdots & & \vdots & \vdots & \vdots \\ 0 & \cdots 0 & xV & xV & 0 & \cdots & 0 & 0 & \cdots 0 & -xV & -xV & 0 & \cdots & 0 \end{bmatrix}\right.\right\}}\Psi_{(m)}\quad(6.9)$$

$$\underbrace{}_{\left(\frac{m}{2}\right)\text{ columns}}\quad\underbrace{}_{\left(\frac{m}{2}\right)\text{ columns}}$$

We call the matrix on the right-hand side (RHS) of Equation 6.9 as single-pulse modulation matrix (**SPMM**), which takes care of all possible single-pulse modulations of the inverter output voltage.

Hence,

$$\mathbf{R'}\,\Psi_{(m)} = \mathbf{SPMM}\ \Psi_{(m)} \qquad (6.10)$$

Now following the same mathematical steps presented in Section 4.3.2.1, we get an equation for the output variable $\mathbf{C'}\,\Psi_{(m)}$ similar to Equation 4.48 as

$$\mathbf{C'}\,\Psi_{(m)} = \frac{1}{m}\left\{\mathbf{SPMM}\right\}\mathbf{W}\left\{\mathbf{WOTF}\right\}\mathbf{W}\ \Psi_{(m)} \qquad (6.11)$$

where:
   WOTF is the Walsh operational transfer function of the load system connected to the inverter output

$\mathbf{C'}\,\Psi_{(m)}$ in Equation 6.11 gives the complete solution in time domain, and each row of $\mathbf{C'}\,\Psi_{(m)}$ gives the solution for a particular value of x and n representing a typical single-pulse modulation.

6.1.1.2 Computation of Normalized Average and rms Load Currents for Single–Pulse Modulation

We consider an inductive load having load parameters $R_L = 4.35\ \Omega$ and $L_L = 0.02$ H.

Considering the normalized load current as the output variable and assuming a time period $T = 0.02$ seconds, the **WOTF** of the load in the scaled Walsh domain is

$$\textbf{WOTF5} = R_\text{L} \left[R_\text{L} \textbf{I} + \left(\frac{\textbf{D}}{\lambda} \right) L_\text{L} \right]^{-1} \tag{6.12}$$

where:

the scaling constant $\lambda = 0.02$

$\textbf{I}$ is an identity matrix

$\textbf{D}$ is the Walsh operational matrix for differentiation

Substituting the respective values of R_L, L_L, and λ, we have

$$\textbf{WOTF5} = 4.35\,[4.35\textbf{I} + \textbf{D}]^{-1} \tag{6.13}$$

For a specific value of m, the unit matrix $\textbf{I}$ and the operational matrix $\textbf{D}$ are known from Equation 3.20.

If we consider $m = 16$, $\textbf{SPMM}$ for this case will be

$$\textbf{SPMM} = \begin{bmatrix} x & 1 & 1 & 1 & 1 & 1 & 1 & x & -x & -1 & -1 & -1 & -1 & -1 & -1 & -x \\ 0 & x & 1 & 1 & 1 & 1 & x & 0 & 0 & -x & -1 & -1 & -1 & -1 & -x & 0 \\ 0 & 0 & x & 1 & 1 & x & 0 & 0 & 0 & 0 & -x & -1 & -1 & -x & 0 & 0 \\ 0 & 0 & 0 & x & x & 0 & 0 & 0 & 0 & 0 & 0 & -x & -x & 0 & 0 & 0 \end{bmatrix}_{(4 \times 16)} \tag{6.14}$$

Since we are interested in the normalized output, all the elements of the $\textbf{SPMM}$ have been divided by the pulse amplitude V.

For $m = 16$, we know the Walsh matrix $\textbf{W}$. Hence, substitution of $\textbf{W}$ and $\textbf{SPMM}$ from Equation 6.14 and $\textbf{WOTF}$ ($=\textbf{WOTF5}$ in this case) from Equation 6.13 in Equation 6.11 gives the solution for the normalized load current. For computation purpose, five different values of x are chosen, namely, $x = 0.2$, 0.4, 0.6, 0.8, and 1.0.

Let the computed output matrix $\textbf{C}'\,\Psi_{(16)}$ be given by

$$\textbf{C}' = \begin{bmatrix} c'_{10} & c'_{11} & c'_{12} & \cdots & c'_{1(n-1)} & c'_{1n} \\ c'_{20} & c'_{21} & c'_{22} & \cdots & c'_{2(n-1)} & c'_{2n} \\ c'_{30} & c'_{31} & c'_{32} & \cdots & c'_{3(n-1)} & c'_{3n} \\ c'_{40} & c'_{41} & c'_{42} & \cdots & c'_{4(n-1)} & c'_{4n} \end{bmatrix}_{(4 \times 16)} \tag{6.15}$$

where:

c'_{in}'s are the coefficients of the output block pulse functions, that is, $i = 1, 2,$ 3, 4 and $n = 0, 1, 2,..., 15$.

To find out the average and root mean square (rms) values of the output variable i_L, we, as earlier, employ the following simple formulas:

The average value is

$$I_{\text{L(av)}} = \frac{1}{16} \sum_{n=0}^{15} c'_{in} \tag{6.16}$$

and the corresponding rms value is

$$I_{L(rms)} = \sqrt{\frac{1}{16} \sum_{n=0}^{15} c'^{\,2}_{in}}$$ (6.17)

for any particular value of i.

The results of the computations are tabulated in Table 6.1, where Walsh domain solutions for the normalized average load currents are compared with the exact solutions. For obtaining the exact solutions, we have employed Equation 6.6, where t_1 is related to x by the following relation:

$$t_1 = (i - x)\frac{T}{m} \text{ seconds}$$ (6.18)

where:
  i denotes the ith row of **SPMM**

It is observed from Table 6.1 that the Walsh domain solutions are remarkably close to the exact solutions, and the maximum percentage error is only 0.660%. Table 6.2 tabulates the rms values of the normalized load current obtained via Walsh analysis. The exact solutions for rms currents could have been found from Equation 6.4, but this being much involved, the effort has been spared. However, from the closeness of the average values in Table 6.1, it is pertinently expected that the rms values will also be in order (Program B.28 in Appendix B).

TABLE 6.1

Normalized Average Load Current for Single-Pulse Modulation ($R_L = 4.35\ \Omega$, $L_L = 0.02\ \text{H}, f = 50\ \text{Hz}, m = 16$)

Normalized Time $t_1\left(= \dfrac{t_1}{T}\right)$	Equivalent Angle (°)	Normalized Average Load Current $I_{L(av)}$		Error (%)
		Walsh Analysis	**Exact**	
0.000	0.0	0.5917	0.5925	0.135
0.0125	4.5	0.5670	0.5697	0.474
0.025	9.0	0.5422	0.5458	0.660
0.0375	13.5	0.5175	0.5209	0.653
0.050	18.0	0.4927	0.4950	0.465
0.075	27.0	0.4390	0.4406	0.363
0.100	36.0	0.3810	0.3831	0.548
0.125	45.0	0.3230	0.3230	0.000
0.1375	49.5	0.2913	0.2922	0.310
0.150	54.0	0.2597	0.2609	0.460
0.1625	58.5	0.2280	0.2292	0.524
0.175	63.0	0.1963	0.1971	0.406
0.200	72.0	0.1317	0.1321	0.303
0.225	81.0	0.0658	0.0662	0.604

TABLE 6.2

Normalized rms Load Current for Single-Pulse Modulation: Walsh Analysis
($R_L = 4.35\ \Omega$, $L_L = 0.02$ H, $f = 50$ Hz, $m = 16$)

Normalized Time $t_1\left(=\dfrac{t_1}{T}\right)$	Equivalent Angle (°)	Normalized rms Load Current $I_{L(rms)}$ Walsh Analysis
0.000	0.0	0.6407
0.0125	4.5	0.6197
0.025	9.0	0.5993
0.0375	13.5	0.5793
0.050	18.0	0.5599
0.075	27.0	0.5118
0.100	36.0	0.4547
0.125	45.0	0.4001
0.1375	49.5	0.3632
0.150	54.0	0.3257
0.1625	58.5	0.2905
0.175	63.0	0.2548
0.200	72.0	0.1759
0.225	81.0	0.0879

6.2 Analysis of an RL Load Fed from a Typical Three-Phase Inverter Line-to-Neutral Voltage

As a further illustration of the Walsh domain operational method, we consider a typical load voltage waveform that may be obtained as one of the line-to-neutral voltages of a three-phase inverter. The waveform is shown in Figure 6.6a. It may be observed that this waveform is not convenient for Walsh analysis because it does not maintain a constant amplitude over each subinterval of (T/m)-second durations, when m subintervals are considered in one period of T seconds.

To make it suitable for Walsh analysis, we use the area-preserving technique and transform the waveform of Figure 6.6a to that of Figure 6.6b. The transformed wave could now very easily be represented by block pulse functions simply by inspection. This has been shown for $m = 16$, and we have the input load voltage $v_o(t)$ as

$$v_o(t) \approx \mathbf{R3'}\ \Psi_{(16)} = V\left[\frac{1}{3}\ \frac{1}{3}\ \frac{4}{9}\ \frac{2}{3}\ \frac{2}{3}\ \frac{4}{9}\ \frac{1}{3}\ \frac{1}{3}\right.$$

$$\left. -\frac{1}{3}\ -\frac{1}{3}\ -\frac{4}{9}\ -\frac{2}{3}\ -\frac{2}{3}\ -\frac{4}{9}\ -\frac{1}{3}\ -\frac{1}{3}\right]\Psi_{(16)} \tag{6.19}$$

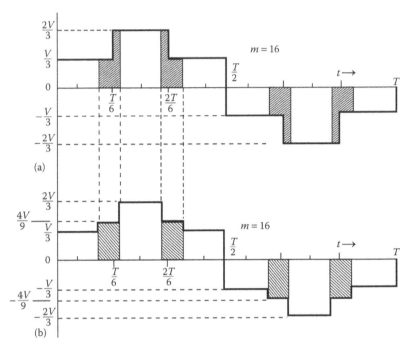

FIGURE 6.6
Line-to-neutral voltage waveform of a three phase inverter (a) before and (b) after area-preserving transformation. (Reprinted from Deb, A. and Datta, A. K., *Int. J. Electron.*, 79, 885–895, 1995. With permission.)

However, for generality, let us assume that for the waveform shown in Figure 6.6a, we consider m number of subintervals in the Walsh domain. Then, the waveform, when converted to the form similar to that shown in Figure 6.6b, will be represented by the following general equation:

$$
\mathbf{R3'}\,\Psi_{(m)} = V\Bigg[\underbrace{\frac{1}{3} \quad \frac{1}{3} \quad \cdots \quad \frac{1}{3}}_{q \text{ terms}} \quad \frac{1}{3}\!\left(\frac{m}{3}-q-q'\right) \underbrace{\frac{2}{3} \quad \frac{2}{3} \quad \cdots \quad \frac{2}{3}}_{q' \text{ terms}}
$$

$$
\frac{1}{3}\!\left(\frac{m}{3}-q-q'\right) \underbrace{\frac{1}{3} \quad \frac{1}{3} \quad \cdots \quad \frac{1}{3}}_{q \text{ terms}} \underbrace{-\frac{1}{3} \quad -\frac{1}{3} \quad \cdots \quad -\frac{1}{3}}_{q \text{ terms}}
$$

$$
-\frac{1}{3}\!\left(\frac{m}{3}-q-q'\right) \underbrace{-\frac{2}{3} \quad -\frac{2}{3} \quad \cdots \quad -\frac{2}{3} \quad -\frac{2}{3}}_{q' \text{ terms}} -\frac{1}{3}\!\left(\frac{m}{3}-q-q'\right)
$$

$$
\underbrace{-\frac{1}{3} \quad -\frac{1}{3} \quad \cdots \quad -\frac{1}{3} \quad -\frac{1}{3}}_{q \text{ terms}}\Bigg]\Psi_{(m)}
$$

(6.20)

where

$$q = \frac{1}{6}[2^p - 3 - (-1)^p]$$

and

$$q' = \frac{1}{2}[2q - 1 + (-1)^p]$$

(6.21)

In the above equation pair, p is any positive integer from 2 to ∞, and m is the number of subintervals in one time period T given by $m = 2^p$.

Inspection of the general equation (6.20) reveals that the total number of terms of the row matrix on its RHS is $(4q + 2q' + 4)$ that is equal to m—the total number of subintervals. Also, for $p = 1$ and $p = 2$, that is, for $m = 2$ and $m = 4$, we get exactly the same wave shape. This is why we have considered the lowest value of p to be equal to 2 in determining q and q' in Equation 6.21. It may also be noted that q as well as q' cannot be negative and their minimum possible value is 0.

To check the validity of Equation 6.20, let us take $m = 16$, that is, $p = 4$. Then from the empirical relations 6.21, we have $q = 2$ and $q' = 2$. Substitution of m, q, and q' in Equation 6.20 yields Equation 6.19 to prove the generality of Equation 6.20.

Now using Equations 6.19 and 3.44—which are similar to Equations 4.43 and 6.11—we can write down the expression for normalized output as

$$\mathbf{C3'}\,\mathbf{\Psi}_{(16)} = \left(\frac{1}{16}\right)\left[\frac{1}{3} \quad \frac{1}{3} \quad \frac{4}{9} \quad \cdots \quad \frac{4}{9} \quad \frac{1}{3} \quad \frac{1}{3} \quad -\frac{1}{3} \quad -\frac{1}{3} \quad -\frac{4}{9} \right.$$

$$\left. \cdots \quad -\frac{4}{9} \quad -\frac{1}{3} \quad -\frac{1}{3}\right]\mathbf{W}\,\{\mathbf{WOTF5}\}\,\mathbf{W}\,\mathbf{\Psi}_{(16)}$$

(6.22)

Here, we have considered a time period of 0.02 seconds, and the same load parameters as in Section 6.1.1.2 are used. So knowing $\mathbf{W}$ and $\mathbf{WOTF5}$ from Equation 6.13, the normalized load current could be determined from Equation 6.22. It is evident that the output will be a row matrix having 16 elements, and the values of the normalized average as well as the rms load currents could be computed using Equations 6.16 and 6.17.

To compare the results obtained by Walsh analysis to those determined from the exact equation, we now derive the necessary exact equation.

Considering the waveform shown in Figure 6.6a, we can represent it by the following equation:

$$v_o(t) = \left(\frac{V}{3}\right)\left[u(t) + u\left(t - \frac{T}{6}\right) - u\left(t - \frac{T}{3}\right) - 2u\left(t - \frac{T}{2}\right) - u\left(t - \frac{2T}{3}\right)\right.$$

$$\left. + u\left(t - \frac{5T}{6}\right) + u(t - T)\right]$$

(6.23)

If R_L and L_L are the load parameters, the Laplace transform of the load current is given by

$$I_L(s) = \frac{\frac{V}{3}}{s(R_L + L_L s)}\left[1 + \exp\left(-\frac{Ts}{6}\right) - \exp\left(-\frac{Ts}{3}\right) - 2\exp\left(-\frac{Ts}{2}\right)\right.$$

$$\left. -\exp\left(-\frac{2Ts}{3}\right) + \exp\left(-\frac{5Ts}{6}\right) + \exp(-Ts)\right] \quad (6.24)$$

Taking the inverse transform and simplifying, the normalized load current is given by

$$i_L(t) = \left(\frac{1}{3}\right)\left[1 - \exp\left(-\frac{R_L t}{L_L}\right)\right]\left[u(t) + u\left(t - \frac{T}{6}\right) - u\left(t - \frac{T}{3}\right)\right.$$

$$\left. -2u\left(t - \frac{T}{2}\right) - u\left(t - \frac{2T}{3}\right) + u\left(t - \frac{5T}{6}\right) + u(t - T)\right] \quad (6.25)$$

Hence, the normalized average current is obtained as

$$I_{L(av)} = \frac{2}{3T}\int_0^{\frac{T}{2}} i_L(t)dt \quad (6.26)$$

Substituting $i_L(t)$ from Equation 6.25 in Equation 6.26 and performing integrations within proper limits, we have, after simplification,

$$I_{L(av)} = \left(\frac{2}{3T}\right)\left[\left(\frac{2T}{3}\right) + \left(\frac{L_L}{R_L}\right)\left\{\exp\left(-\frac{R_L T}{2L_L}\right) + \exp\left(-\frac{R_L T}{3L_L}\right) - \exp\left(-\frac{R_L T}{6L_L}\right) - 1\right\}\right] \quad (6.27)$$

For $T = 0.02$ seconds, $R_L = 4.35\ \Omega$, and $L_L = 0.02$ H, we have from the above equation

$$I_{L(av)} = 0.2703 \text{ (exact)}$$

From Equations 6.16 and 6.22, the normalized average load current is obtained as

$$I_{L(av)} = 0.2697 \text{ (Walsh)}$$

which is found to be very close to the exact value (Program B.29 in Appendix B).

As in the earlier cases, we have not computed the exact values of the normalized rms current due to many involved mathematical operations, and only Walsh domain solution is obtained from Equations 6.17 and 6.22, which is given below:

$$I_{L(rms)} = 0.3018 \text{ (Walsh)}$$

However, Equation 6.25 could be employed effectively to determine the exact value of the normalized rms current if so desired.

From the average values obtained, it is noted that the percentage error in the Walsh domain solution is only 0.222%.

6.3 Conclusion

In this chapter, the Walsh domain operational method has been applied to analyze inverter circuits. In this analysis, it has been assumed that all the semiconductor devices act as ideal switches and commutation time is negligibly small. This will help in producing the ideal waveforms presented in the analysis.

Another point to mention is the switching transients that are readily obtained in the solution of load current. This may be looked upon as an advantage of the proposed method, since it is rather tricky to obtain the transient solution of such a PE system that handles complicated voltage waveforms. However, to obtain the steady-state solution, the Walsh method has no other way but to consider a few load voltage cycles instead of only one cycle. This will make the transient insignificant with respect to the steady-state solution. Such a solution will require a higher value of m, say 64, 128, or 256. When working with higher values of m, the added advantage would be the improved accuracy of the Walsh analysis.

References

1. Bedford, B. D. and Hoft, R. G., *Principles of Inverter Circuits*, John Wiley & Sons, New York, 1964.
2. Slonim, M. A., *Theory of Static Converter Systems*, Elsevier, Amsterdam, 1984.
3. Sen, P. C., *Power Electronics*, Tata McGraw-Hill Publishing, New Delhi, 1987.
4. Heumann, K., *Fundamentals of Power Electronics*, AEG–Telefunken, Frankfurt am Main, 1979.
5. Nonaka, S. and Oyama, J., Characteristics of three-phase, salient-synchronous motors driven by thyristor inverters, *Electr. Eng. Jpn.*, Vol. **92**, No. 3, pp. 85–94, 1972.
6. Taniguchi, K., Irie, H. and Ishizuki, T., Pulsewidth-modulated power amplifiers consisting of thyristor-bridge inverters, *Electr. Eng. Jpn.*, Vol. **93**, No. 5, pp. 385–390, 1973.
7. Kazmierkowski, M. and Malesni, L., Current control techniques for three phase voltage-source PWM converter: A survey, *IEEE Trans. Ind. Electron.*, Vol. **45**, No. 5, pp. 691–703, 1998.

8. Rahman, M., Radwin, T., Osheiba, A. and Lashine, A., Analysis of current controllers for voltage source inverter, *IEEE Trans. Ind. Electron.*, Vol. **44**, No. 4, pp. 477–485, 1997.

9. Bose, B. K. (Ed.), *Adjustable Speed AC Drive Systems*, IEEE Press, New York, 1981.

10. Deb, A. and Datta, A. K., Analysis of pulse-fed power electronic circuits using Walsh functions, *Int. J. Electron.*, Vol. **62**, No. 3, pp. 449–459, 1987.

11. Deb, A. and Datta, A. K., Analysis of a continuously variable pulse-width modulated system via Walsh functions, *Int. J. Syst. Sci.*, Vol. **23**, No. 2, pp. 151–166, 1992.

12. Deb, A. and Datta, A. K., On Walsh/block pulse domain analysis of power electronic circuits: Part 2—Continuously pulse-width modulated inverter, *Int. J. Electron.*, Vol. **79**, No. 6, pp. 885–895, 1995.

13. Dewan, S. B. and Straughen, A., *Power Semiconductor Circuits*, John Wiley & Sons, New York, 1975.

14. Dubey, G. K., *Fundamentals of Electric Drives* (2nd Ed.), Narosa Publishing House, Kolkata, 2001.

Appendix A: Introduction to Linear Algebra

A matrix is a rectangular array of variables, mathematical expressions, or simply numbers. Commonly, a matrix is written as

$$
\mathbf{A} = \begin{bmatrix}
a_{11} & a_{12} & \cdots & a_{1n} \\
a_{21} & a_{22} & \cdots & a_{2n} \\
\vdots & \vdots & \ddots & \vdots \\
a_{m1} & a_{m2} & \cdots & a_{mn}
\end{bmatrix}
\tag{A.1}
$$

The size of the matrix, with m rows and n columns, is called an m-by-n (or $m \times n$) matrix, where m and n are called its dimensions.

A matrix with one row [a $(1 \times n)$ matrix] is called a row vector and a matrix with one column [an $(m \times 1)$ matrix] is called a column vector. Any isolated row or column of a matrix is a row or column vector, obtained by removing all other rows or columns, respectively, from the matrix.

A.1 Square Matrices

A square matrix is a matrix with $m = n$, that is, the same number of rows and columns. An n-by-n matrix is known as a square matrix of order n. Any two square matrices of the same order can be added, subtracted, or multiplied.

For example, each of the following matrices is a square matrix of order 4, with four rows and four columns:

$$
\mathbf{A} = \begin{bmatrix}
1 & 2 & 7 & 0 \\
3 & 4 & 3 & 1 \\
8 & 3 & 1 & 1 \\
2 & 4 & 0 & 3
\end{bmatrix}, \mathbf{B} = \begin{bmatrix}
3 & 1 & 0 & -1 \\
2 & -3 & 4 & 2 \\
1 & 0 & 9 & 1 \\
-3 & 2 & -1 & 0
\end{bmatrix}
$$

Then,

$$
\mathbf{A} + \mathbf{B} =
\begin{bmatrix}
4 & 3 & 7 & -1 \\
5 & 1 & 7 & 3 \\
9 & 3 & 10 & 2 \\
-1 & 6 & -1 & 3
\end{bmatrix}
\quad \text{and} \quad
\mathbf{A} - \mathbf{B} =
\begin{bmatrix}
-2 & 1 & 7 & 1 \\
1 & 7 & -1 & -1 \\
7 & 3 & -8 & 0 \\
5 & 2 & 1 & 3
\end{bmatrix}
$$

A.2 Determinant

The determinant [written as det(**A**) or |**A**|] of a square matrix **A** is a number encoding certain properties of the matrix. A matrix is invertible, if and only if its determinant is nonzero.

The determinant of a 2-by-2 matrix is given by

$$
\det\begin{pmatrix} a & b \\ c & d \end{pmatrix} = ad - bc \tag{A.2}
$$

A.2.1 Properties

The determinant of a product of square matrices equals the product of their determinants: $\det(\mathbf{AB}) = \det(\mathbf{A}) \cdot \det(\mathbf{B})$

Adding a multiple of any row to another row, or a multiple of any column to another column, does not change the determinant. Interchanging two rows or two columns affects the determinant by multiplying it by –1.

Using these operations, any matrix can be transformed to a lower (or upper) triangular matrix, and for such matrices, the determinant equals the product of the entries on the main diagonal.

A.2.2 Orthogonal Matrix

An orthogonal matrix is a square matrix with real entries whose columns and rows are orthogonal vectors, that is, orthonormal vectors.

Equivalently, a matrix **A** is orthogonal if its transpose is equal to its inverse. That is,

$$
\mathbf{A}^\mathrm{T} = \mathbf{A}^{-1}
$$

which implies

$$
\mathbf{A}^\mathrm{T} \cdot \mathbf{A} = \mathbf{A} \cdot \mathbf{A}^\mathrm{T} = \mathbf{I}
$$

where:

I is the identity matrix

An orthogonal matrix **A** is necessarily invertible with inverse $\mathbf{A}^{-1} = \mathbf{A}^\mathrm{T}$. The determinant of any orthogonal matrix is either +**I** or –**I**.

A.2.3 Trace of a Matrix

In Equation A.1, the entries $a_{i,i}$ form the main diagonal of the matrix **A**. The trace, tr(**A**), of the square matrix **A** is the sum of its diagonal entries. The trace of the product of two matrices is independent of the order of the factors **A** and **B**. That is,

$$\text{tr}(\mathbf{AB}) = \text{tr}(\mathbf{BA})$$

Also, the trace of a matrix is equal to that of its transpose: $\text{tr}(\mathbf{A}) = \text{tr}(\mathbf{A}^T)$.

A.2.4 Diagonal, Lower Triangular, and Upper Triangular Matrices

If all entries of a matrix except those of the main diagonal are zero, the matrix is called a diagonal matrix. If only all entries above (or below) the main diagonal are zero, it is called a lower triangular matrix (or upper triangular matrix).

For example, a diagonal matrix of order 3 is

$$\begin{bmatrix} d_{11} & 0 & 0 \\ 0 & d_{22} & 0 \\ 0 & 0 & d_{33} \end{bmatrix}$$

A lower triangular matrix of order 3 is

$$\begin{bmatrix} l_{11} & 0 & 0 \\ l_{21} & l_{22} & 0 \\ l_{31} & l_{32} & l_{33} \end{bmatrix}$$

and an upper triangular matrix of similar order is

$$\begin{bmatrix} u_{11} & u_{12} & u_{13} \\ 0 & u_{22} & u_{23} \\ 0 & 0 & u_{33} \end{bmatrix}$$

A.2.5 Symmetric Matrix

A square matrix **A** that is equal to its transpose, that is, $\mathbf{A} = \mathbf{A}^T$, is a symmetric matrix. If, instead, **A** was equal to the negative of its transpose, that is, $\mathbf{A} = -\mathbf{A}^T$, **A** is a skew-symmetric matrix.

A.2.6 Identity Matrix or Unit Matrix

If $\mathbf{A}$ is a square matrix,

$$\mathbf{AI} = \mathbf{IA} = \mathbf{A}$$

where:
  $\mathbf{I}$ is the identity matrix of the same order

The identity matrix $\mathbf{I}_{(n)}$ of size n is the n-by-n matrix in which all the elements on the main diagonal are equal to 1 and all other elements are equal to 0. An identity matrix of order 3 is

$$\mathbf{I}_{(3)} = \begin{bmatrix} 1 & 0 & 0 \\ 0 & 1 & 0 \\ 0 & 0 & 1 \end{bmatrix}$$

It is called identity matrix because multiplication with it leaves a matrix unchanged. If $\mathbf{A}$ is an $(m \times n)$ matrix, then

$$\mathbf{A}_{(m \times n)}\mathbf{I}_{(n)} = \mathbf{I}_{(m)}\mathbf{A}_{(m \times n)}$$

A.2.7 Transpose of a Matrix

The transpose $\mathbf{A}^T$ of a square matrix $\mathbf{A}$ can be obtained by reflecting the elements along its main diagonal. Repeating the process on the transposed matrix returns the elements to their original position.

The transpose of a matrix may be obtained by any one of the following equivalent actions:

1. Reflect $\mathbf{A}$ over its main diagonal to obtain $\mathbf{A}^T$.
2. Write the rows of $\mathbf{A}$ as the columns of $\mathbf{A}^T$.
3. Write the columns of $\mathbf{A}$ as the rows of $\mathbf{A}^T$.

Formally, the ith row, jth column element of $\mathbf{A}^T$ is the jth row, ith column element of $\mathbf{A}$. That is,

$$[\mathbf{A}^T]_{ij} = \mathbf{A}_{ji}$$

If $\mathbf{A}$ is an $(m \times n)$ matrix, $\mathbf{A}^T$ is an $(n \times m)$ matrix.

A.2.8 Properties

For matrices $\mathbf{A}$, $\mathbf{B}$, and scalar c, we have the following properties of transpose:

1. $(\mathbf{A}^{\mathsf{T}})^{\mathsf{T}} = \mathbf{A}$
2. $(\mathbf{A} + \mathbf{B})^{\mathsf{T}} = \mathbf{A}^{\mathsf{T}} + \mathbf{B}^{\mathsf{T}}$
3. $(\mathbf{AB})^{\mathsf{T}} = \mathbf{B}^{\mathsf{T}} \mathbf{A}^{\mathsf{T}}$

   Note that the order of the factors above reverses. From this, one can deduce that a square matrix $\mathbf{A}$ is invertible if and only if $\mathbf{A}^{\mathsf{T}}$ is invertible, and in this case, we have $(\mathbf{A}^{-1})^{\mathsf{T}} = (\mathbf{A}^{\mathsf{T}})^{-1}$. By induction, this result extends to the general case of multiple matrices, where we find that

   $$(\mathbf{A}_1\mathbf{A}_2...\mathbf{A}_{k-1}\mathbf{A}_k)^{\mathsf{T}} = \mathbf{A}_k^{\mathsf{T}}\mathbf{A}_{k-1}^{\mathsf{T}}...\mathbf{A}_2^{\mathsf{T}}\mathbf{A}_1^{\mathsf{T}}$$

4. $(c\mathbf{A})^{\mathsf{T}} = c\mathbf{A}^{\mathsf{T}}$

   The transpose of a scalar is the same scalar.
5. $\det(\mathbf{A}^{\mathsf{T}}) = \det(\mathbf{A})$

A.3 Matrix Multiplication

Matrix multiplication is a binary operation that takes a pair of matrices and produces another matrix. This term normally refers to the matrix product.

Multiplication of two matrices is defined only if the number of columns of the left matrix is the same as the number of rows of the right matrix. If $\mathbf{A}$ is an m-by-n matrix and $\mathbf{B}$ is an n-by-p matrix, their matrix product $\mathbf{AB}$ is the m-by-p matrix whose entries are given by dot product of the corresponding row of $\mathbf{A}$ and the corresponding column of $\mathbf{B}$. That is,

$$[\mathbf{AB}]_{i,j} = A_{i,1}B_{1,j} + A_{i,2}B_{2,j} + \cdots + A_{i,n}B_{n,j} = \sum_{r=1}^{n} A_{i,r}B_{r,j}$$

where:
$1 \leq i \leq m$
$1 \leq j \leq p$

Matrix multiplication satisfies the following rules:

1. $(\mathbf{AB})\mathbf{C} = \mathbf{A}(\mathbf{BC})$ (associativity)
2. $(\mathbf{A} + \mathbf{B})\mathbf{C} = \mathbf{AC} + \mathbf{BC}$
3. $\mathbf{C}(\mathbf{A} + \mathbf{B}) = \mathbf{CA} + \mathbf{CB}$ (left and right distributivity)

whenever the size of the matrices is such that the various products are defined.

The product **AB** may be defined without **BA** being defined, namely, if **A** and **B** are *m-by-n* and *n-by-k* matrices, respectively, and $m \neq k$.

Even if both products are defined, they need not be equal. The relationship is

$$\mathbf{AB} \neq \mathbf{BA}$$

it means, matrix multiplication is not commutative, in marked contrast to (rational, real, or complex) numbers whose product is independent of the order of the factors. An example of two matrices not commuting with each other is

$$\begin{bmatrix} 5 & 2 \\ 3 & 3 \end{bmatrix} \begin{bmatrix} 1 & 0 \\ 0 & 0 \end{bmatrix} = \begin{bmatrix} 5 & 0 \\ 3 & 0 \end{bmatrix}$$

whereas

$$\begin{bmatrix} 1 & 0 \\ 0 & 0 \end{bmatrix} \begin{bmatrix} 5 & 2 \\ 3 & 3 \end{bmatrix} = \begin{bmatrix} 5 & 2 \\ 0 & 0 \end{bmatrix}$$

Since det(**A**) and det(**B**) are just numbers and so commute, det(**AB**) = det(**A**) det(**B**) = det(**B**)det(**A**) = det(**BA**), even when **AB** $\neq$ **BA**.

A.3.1 A Few Properties of Matrix Multiplication

Matrix multiplication have the following properties:

1. Associative

   $\mathbf{A}(\mathbf{BC}) = (\mathbf{AB})\mathbf{C}$

2. Distributive over matrix addition

   $\mathbf{A}(\mathbf{B} + \mathbf{C}) = \mathbf{AB} + \mathbf{AC}$ and $(\mathbf{A} + \mathbf{B})\mathbf{C} = \mathbf{AC} + \mathbf{BC}$

3. Scalar multiplication is compatible with matrix multiplication

   $\lambda(\mathbf{AB}) = (\lambda\mathbf{A})\mathbf{B}$ and $(\mathbf{AB})\lambda = \mathbf{A}(\mathbf{B}\lambda)$

where:
  λ is a scalar

If the entries of the matrices are real or complex numbers, all four quantities are equal.

A.3.2 Inverse of a Matrix

If **A** is a square matrix, there may be an inverse matrix $\mathbf{A}^{-1} = \mathbf{B}$ such that

$$\mathbf{AB} = \mathbf{BA} = \mathbf{I}$$

If this property holds, **A** is an invertible matrix. If not, **A** is a singular or degenerate matrix. A singular matrix has a determinant equal to zero.

A.3.3 Analytic Solution of the Inverse

Inverse of a square nonsingular matrix **A** may be computed by writing the transpose of the matrix **C** of cofactors of **A**, known as an adjoint matrix. The matrix **C** is divided by the determinant of **A** to find **A**⁻¹. That is,

$$\mathbf{A}^{-1} = \frac{1}{\det(\mathbf{A})}(\mathbf{C}^{\mathsf{T}})_{ij} = \frac{1}{\det(\mathbf{A})}(\mathbf{C}_{ji}) = \frac{1}{\det(\mathbf{A})}\begin{pmatrix} C_{11} & C_{21} & \cdots & C_{n1} \\ C_{12} & C_{22} & \cdots & C_{n2} \\ \vdots & \vdots & \ddots & \vdots \\ C_{1n} & C_{2n} & \cdots & C_{nn} \end{pmatrix} \quad \text{(A.3)}$$

where:
  $\det(\mathbf{A})$ is the determinant of **A**
  $\mathbf{C}_{ij}$ is the matrix of cofactors
  $\mathbf{C}^{\mathsf{T}}$ denotes the transpose of **C**

A.3.4 Inversion of a 2 × 2 Matrix

The cofactor equation listed above yields the following result for a 2 × 2 matrix. Let the matrix to be inverted be

$$\mathbf{A} = \begin{bmatrix} a & b \\ c & d \end{bmatrix}$$

Then, its inverse is given by

$$\mathbf{A}^{-1} = \begin{bmatrix} a & b \\ c & d \end{bmatrix}^{-1} = \frac{1}{\det(\mathbf{A})}\begin{bmatrix} d & -b \\ -c & a \end{bmatrix} = \frac{1}{(ad-bc)}\begin{bmatrix} d & -b \\ -c & a \end{bmatrix}$$

using Equations A.2 and A.3.

A.3.5 Inversion of a 3 × 3 Matrix

Let the matrix to be inverted be

$$\mathbf{A} = \begin{bmatrix} a & b & c \\ d & e & f \\ g & h & k \end{bmatrix}$$

Then, its inverse is given by

$$
\mathbf{A}^{-1} = \begin{bmatrix} a & b & c \\ d & e & f \\ g & h & k \end{bmatrix}^{-1} = \frac{1}{\det(\mathbf{A})} \begin{bmatrix} A & B & C \\ D & E & F \\ G & H & K \end{bmatrix}^{T} = \frac{1}{\det(\mathbf{A})} \begin{bmatrix} A & D & G \\ B & E & H \\ C & F & K \end{bmatrix} \quad \text{(A.4)}
$$

where:
   $A, B, C, D, E, F, G, H,$ and K are the cofactors of the matrix $\mathbf{A}$

The determinant of $\mathbf{A}$ can be computed as follows:

$$
\det(\mathbf{A}) = a(ek - fh) - b(kd - fg) + c(dh - eg)
$$

If the determinant is nonzero, the matrix is invertible, with the cofactors given by

$$
\begin{aligned}
A &= (ek - fh) & D &= (ch - bk) & G &= (bf - ce) \\
B &= (fg - dk) & E &= (ak - cg) & H &= (cd - af) \\
C &= (dh - eg) & F &= (gb - ah) & K &= (ae - bd)
\end{aligned}
$$

Determination of these cofactors subsequently leads to the computation of the inverse of $\mathbf{A}$.

A.4 Similarity Transformation

Two n-by-n matrices $\mathbf{A}$ and $\mathbf{B}$ are called similar if

$$
\mathbf{B} = \mathbf{P}^{-1}\mathbf{A}\mathbf{P} \tag{A.5}
$$

for some invertible n-by-n matrix $\mathbf{P}$.
   Similar matrices represent the same linear transformation under two different bases, with $\mathbf{P}$ being the change of basis matrix.
   The determinant of the similarity transformation of a matrix is equal to the determinant of the original matrix $\mathbf{A}$.

$$
\det(\mathbf{B}) = \det(\mathbf{P}^{-1}\mathbf{A}\mathbf{P}) = \det(\mathbf{P}^{-1})\det(\mathbf{A})\det(\mathbf{P}) = \frac{\det(\mathbf{A})}{\det(\mathbf{P})}\det(\mathbf{P}) = \det(\mathbf{A}) \quad \text{(A.6)}
$$

Also, the eigenvalues of the matrices **A** and **B** are also the same. That is,

$$
\begin{aligned}
\det(\mathbf{B} - \lambda\mathbf{I}) &= \det(\mathbf{P}^{-1}\mathbf{A}\,\mathbf{P} - \lambda\mathbf{I}) \\
&= \det(\mathbf{P}^{-1}\mathbf{A}\,\mathbf{P} - \mathbf{P}^{-1}\lambda\mathbf{I}\,\mathbf{P}) \\
&= \det(\mathbf{P}^{-1}(\mathbf{A} - \lambda\mathbf{I})\mathbf{P}) \\
&= \det(\mathbf{P}^{-1})\det(\mathbf{A} - \lambda\mathbf{I})\det(\mathbf{P}) \\
&= \det(\mathbf{A} - \lambda\mathbf{I})
\end{aligned}
\tag{A.7}
$$

where:
  λ is a scalar

Appendix B: Selected MATLAB® Programs

B.1 Program for Representation of a Function $f_1(t) = \sin(\pi t)$ Using Walsh/BPF Function (Examples 2.1 and 2.3)

```
clc
clear all

%%- - - - - - - - - - EXACT SOLUTION - - - - - - - - - - - - -%%

k=0.01
for i=1:101
    t(i)=(i-1)*k
    FUNC(i)=sin(pi*(i-1)*k);
end

        %%- - - - - - FUNCTION PLOTTING- - - - - - %%

plot(t,FUNC,'k')                % plot the exact function
hold on

%%- - - - - - - - - - SOLUTION VIA BPF - - - - - - - - - - - -%%

clear all
clc
syms t

func=sin(pi*t)                  % the function to be approximated
T=1                             % total time considered
m=8                             % no. of Walsh components
h=T/m                           % length of each interval
for i=1:m
     x1(i)=double(m*int(func,((i-1)*h),(i*h)));

     bpfcoeff(1,i)=x1(i); % BPF coefficients of the function
end

W=[1  1  1  1  1  1  1  1;
   1  1  1  1 -1 -1 -1 -1;
   1  1 -1 -1  1  1 -1 -1;
   1  1 -1 -1 -1 -1  1  1;
   1 -1  1 -1  1 -1  1 -1;
   1 -1  1 -1 -1  1 -1  1;
   1 -1 -1  1  1 -1 -1  1;
   1 -1 -1  1 -1  1  1 -1] % Walsh matrix
```

```
Walshcoeff=bpfcoeff*inv(W)          % Walsh coefficents
                                    % of the function

bpfcoeff1=Walshcoeff*W
bpfcoeff=[bpfcoeff bpfcoeff(1,m)]


        %%- - - - - - - - - FUNCTION PLOTTING- - - - - - - - %%

t=0:h:T
hold on
stairs(t,bpfcoeff)   % plot the solution via BPF
```

B.2 Program for Representation of a Function $f_2(t) = \exp(-t)$ Using Walsh/BPF Function (Examples 2.2 and 2.4)

```
clc
clear all

%%- - - - - - - - - - - EXACT SOLUTION - - - - - - - - - - - -%%

k=0.01
for i=1:101
    t(i)=(i-1)*k
    FUNC(i)=exp(-(i-1)*k);
end


        %%- - - - - - FUNCTION PLOTTING- - - - - - %%

plot(t,FUNC,'k')              % plot the exact function

%%- - - - - - - - - - SOLUTION VIA BPF - - - - - - - - - - - -%%

clc
clear all
syms t

func=exp(-t)                 % the function to be approximated
T=1                          % total time considered
m=8                          % no. of Walsh components
h=T/m                        % length of each interval

for i=1:m
    x1(i)=double(m*int(func,((i-1)*h),(i*h)));
    bpfcoeff(1,i)=x1(i);     % BPF coefficients of the function
end
```

```
W=[1  1  1  1  1  1  1  1;
   1  1  1  1 -1 -1 -1 -1;
   1  1 -1 -1  1  1 -1 -1;
   1  1 -1 -1 -1 -1  1  1;
   1 -1  1 -1  1 -1  1 -1;
   1 -1  1 -1 -1  1 -1  1;
   1 -1 -1  1  1 -1 -1  1;
   1 -1 -1  1 -1  1  1 -1]              % Walsh matrix

Walshcoeff=bpfcoeff*inv(W)  % Walsh coefficents of the function

bpfcoeff1=Walshcoeff*W
bpfcoeff=[bpfcoeff bpfcoeff(1,m)]

      %%- - - - - - - - - FUNCTION PLOTTING- - - - - - - - %%

t=0:h:T
hold on
stairs(t,bpfcoeff) % plot the solution via BPF
```

B.3 Program for Representation of a Function $f(t) = t$ Using Walsh/BPF Function for $m = 4$ (Section 3.1.1, Figure 3.2)

```
clc
clear all

%%- - - - - - - - - - - EXACT SOLUTION - - - - - - - - - - -%%

k=0.01
for i=1:101
    t(i)=(i-1)*k
    FUNC(i)=t(i);
end

      %%- - - - - FUNCTION PLOTTING- - - - - %%

plot(t,FUNC,'k')              % plot the exact function

%%- - - - - - - - - - - SOLUTION VIA BPF - - - - - - - - - -%%

clear all
clc
syms t

func=t              % the function to be approximated
T=1                 % total time considered
```

```
m=4                        % no. of Walsh components
h=T/m                      % length of each interval

for i=1:m
    x1(i)=double(m*int(func,((i-1)*h),(i*h)));
    bpfcoeff(1,i)=x1(i);   % BPF coefficients of the function
end

W=[1  1  1  1;
   1  1 -1 -1;
   1 -1  1 -1;
   1 -1 -1  1]             % Walsh matrix

Walshcoeff=bpfcoeff*inv(W)       % Walsh coefficents
                                 % of the function

bpfcoeff=[bpfcoeff bpfcoeff(1,m)]

      %%- - - - - - - - FUNCTION PLOTTING- - - - - - - - %%

t=0:h:T
hold on
stairs(t,bpfcoeff)         % plot the solution via BPF
```

B.4 Program for Representation of a Function $f(t) = t$ Using Walsh/BPF Function for $m = 8$ (Section 3.1.1, Figure 3.4)

```
clc
clear all

%%- - - - - - - - - - EXACT SOLUTION - - - - - - - - - - - -%%

k=0.01
for i=1:101
    t(i)=(i-1)*h
    FUNC(i)=t(i);
end

        %%- - - - - - FUNCTION PLOTTING- - - - - - %%

plot(t,FUNC,'k')              % plot the exact function

%%- - - - - - - - - - SOLUTION VIA BPF - - - - - - - - - - %%

clear all
clc
syms t
```

```
func=t                      % the function to be approximated
T=1                         % total time considered
m=8                         % no. of Walsh components
h=T/m                       % length of each interval

for i=1:m
    x1(i)=double(m*int(func,((i-1)*h),(i*h)));
    bpfcoeff(1,i)=x1(i);    % BPF coefficients of the function
end

W=[1  1  1  1  1  1  1  1;
   1  1  1  1 -1 -1 -1 -1;
   1  1 -1 -1  1  1 -1 -1;
   1  1 -1 -1 -1 -1  1  1;
   1 -1  1 -1  1 -1  1 -1;
   1 -1  1 -1 -1  1 -1  1;
   1 -1 -1  1  1 -1 -1  1;
   1 -1 -1  1 -1  1  1 -1]  % Walsh matrix

Walshcoeff=bpfcoeff*inv(W)        % Walsh coefficents
                                  % of the function
bpfcoeff=[bpfcoeff bpfcoeff(1,m)]


        %%- - - - - - - - - FUNCTION PLOTTING- - - - - - - - %%

t=0:h:T

hold on
stairs(t,bpfcoeff)          % plot the solution via BPF
```

B.5 Program for Integration of a Function $f_1(t) = \sin(\pi t)$ Using Walsh/BPF Function for $m = 4$ (Example 3.1)

```
clc
clear all

%%- - - - - - - - - - - EXACT SOLUTION - - - - - - - - - - - -%%

syms t t1

FUNC1=int(sin(pi*t),0,t1)       % integration of f(t)=sin(pi*t)
k=0.01;
for i=1:101
    t2(i)=(i-1)*k;
    FUNC2(i)=(1-cos(pi*t2(i)))/pi;
end
```

```
%%- - - - - - FUNCTION PLOTTING- - - - - - %%

plot(t2,FUNC2,'k')              % plot the exact function

%%- - - - - - SOLUTION VIA DIRECT EXPANSION - - - - - - - -%%

func=sin(pi*t);                 % given function
T=1                             % total time considered
m=4                             % no. of Walsh components
h=T/m                           % length of each interval

W=[1  1   1   1;
   1  1  -1  -1;
   1 -1   1  -1;
   1 -1  -1   1];               % Walsh matrix

func1=FUNC1;                    % integration of the given function

for i=1:m
    x1(i)=double(m*int(func1,((i-1)*h),(i*h)));
    bpfcoeff1(1,i)=x1(i);  % BPF coefficients of integrated
                           % function via direct expansion
end

Walshcoeff1=bpfcoeff1*inv(W)            % Walsh coefficents of
                                       % the given function
                                       % via direct expansion

%%- - - SOLUTION VIA OPERATIONAL MATRIX FOR INTEGRATION - - -%%

opnmatrix_walsh=[1/2 -1/4  -1/8     0;
                 1/4    0     0  -1/8;
                 1/8    0     0     0;
                   0  1/8     0     0];
                            % operational matrix for
                            % integration in Walsh domain

for i=1:m
    x1(i)=double(m*int(func,((i-1)*h),(i*h)));
    bpfcoeff(1,i)=x1(i);            % BPF coefficients of given
                                    % function
end

Walshcoeff=bpfcoeff*inv(W);         % Walsh coefficents of the
                                    % given function
Walshcoeff_int=Walshcoeff*opnmatrix_walsh
            % Walsh coefficents of the integrated function
                                    % via operational matrix for
                                    % integration in Walsh domain
```

```
bpfcoeff_int=Walshcoeff_int*W;    % BPF coefficents of the
                                  % integrated function
                                  % via operational matrix for
                                  % integration in Walsh domain


        %%- - - - - - FUNCTION PLOTTING- - - - - - %%


t=0:h:T;
hold on
stairs(t,bpfcoeff1,'b')    % plot the Walsh coefficients via
                           % direct expansion
hold on
stairs(t,[bpfcoeff_int 0],'r')    % plot the Walsh coefficents
                                  % via operational
                                  % matrix for integration
hold on


%%- - - - PERCENTAGE ERROR BETWEEN WALSH COEFFICIENTS OBTAINED
% VIA DIRECT EXPANSION AND OPERATIONAL MATRIX FOR INTEGRATION %%


for i=1:m
    percentageerror(i)=((Walshcoeff1(i)-Walshcoeff_
    int(i))*100)/Walshcoeff1(i);
end
percentageerror % percentage error between Walsh coefficients
                % obtained via direct expansion and operational
                % matrix for integration


        %%- - - - - - - MISE Calculation- - - - - - -%%


miseerrorbpf=zeros(1,1);
for j=1:m
    error1(1,j)=double(int((func1-bpfcoeff1(1,j))^2,
    ((j-1)*h),(j*h)));
    miseerrorbpf=miseerrorbpf+error1(1,j);
end
miseerrorbpf          % MISE for the coefficients
                      % obtained via direct expansion
miseerrorbpf_int=zeros(1,1);
for j=1:m
    error2(1,j)=double(int((func1-bpfcoeff_int(1,j))^2,
    ((j-1)*h),(j*h)));
    miseerrorbpf_int=miseerrorbpf_int+error2(1,j);
end
miseerrorbpf_int      % MISE for the coefficients obtained via
                      % operational matrix for integration
```

B.6 Program for Integration of a Function $f_1(t) = \sin(\pi t)$ Using Walsh/BPF Function for $m = 8$ (Example 3.1)

```
clc
clear all

%%- - - - - - - - - - - EXACT SOLUTION- - - - - - - - - - - - -%%

syms t t1

FUNC1=int(sin(pi*t),0,t1)         % integration of f(t)=sin(pi*t)

k=0.01;
for i=1:101
    t2(i)=(i-1)*k;
    FUNC2(i)=(1-cos(pi*t2(i)))/pi;
end

        %%- - - - - - FUNCTION PLOTTING- - - - - - %%

plot(t2,FUNC2,'k')                    % plot the exact function

%%- - - - - - - SOLUTION VIA DIRECT EXPANSION - - - - - - - -%%

func=sin(pi*t);                % given function
T=1                            % total time considered
m=8                            % no. of Walsh components
h=T/m                          % length of each interval

W=[1   1   1   1   1   1   1   1;
   1   1   1   1 -1 -1 -1 -1;
   1   1 -1 -1   1   1 -1 -1;
   1   1 -1 -1 -1 -1   1   1;
   1 -1   1 -1   1 -1   1 -1;
   1 -1   1 -1 -1   1 -1   1;
   1 -1 -1   1   1 -1 -1   1;
   1 -1 -1   1 -1   1   1 -1]  % Walsh matrix

func1=FUNC1;        % integration of the given function

for i=1:m
    x1(i)=double(m*int(func1,((i-1)*h),(i*h)));
    bpfcoeff1(1,i)=x1(i);  %BPF coefficients of integrated
                           % function via direct expansion
end

Walshcoeff1=bpfcoeff1*inv(W)    % Walsh coefficents of the
                                % intregated
                                % function via direct expansion
```

```
%%- - - SOLUTION VIA OPERATIONAL MATRIX FOR INTEGRATION - - -%%

opnmatrix_walsh=[1/2  -1/4  -1/8     0 -1/16     0     0    0;
                 1/4     0     0  -1/8     0 -1/16     0    0;
                 1/8     0     0     0     0     0 -1/16    0;
                   0   1/8     0     0     0     0     0 -1/16;
                1/16     0     0     0     0     0     0    0;
                   0  1/16     0     0     0     0     0    0;
                   0     0  1/16     0     0     0     0    0;
                   0     0     0  1/16     0     0     0    0]
```

    % operational matrix for integration in Walsh domain

```
for i=1:m
    x1(i)=double(m*int(func,((i-1)*h),(i*h)));
    bpfcoeff(1,i)=x1(i);    %BPF coefficients of given function
end
```

```
Walshcoeff=bpfcoeff*inv(W);              % Walsh coefficents of
                                         % the given function
```

```
Walshcoeff_int=Walshcoeff*opnmatrix_walsh
                        % Walsh coefficents of the integrated
                        % function via operational matrix for
                        % integration in Walsh domain
```

```
bpfcoeff_int=Walshcoeff_int*W;
                        % BPF coefficents of the integrated
                        % function via operational matrix for
                        % integration in Walsh domain
```

```
    %%- - - - - - FUNCTION PLOTTING- - - - - - %%

t=0:h:T;
hold on
stairs(t,[bpfcoeff1 0],'b')    % plot the Walsh coefficients
                               % via direct expansion
hold on
stairs(t,[bpfcoeff_int 0],'r')    % plot the Walsh coefficients
                                  % via operational
                                  % matrix for integration
hold on
```

```
%%- - - - PERCENTAGE ERROR BETWEEN WALSH COEFFICIENTS OBTAINED
%- - - VIA DIRECT EXPANSION AND OPERATIONAL MATRIX FOR
INTEGRATION- - - %%

for i=1:m
    percentageerror(i)=((Walshcoeff1(i)-Walshcoeff_int(i))*100)/
    Walshcoeff1(i);
end
```

```
percentageerror              % percentage error between Walsh
                             % coefficients obtained via direct
                             % expansion and operational matrix
                             % for integration

        %%- - - - - - - MISE CALCULATION- - - - - - -%%

miseerrorbpf=zeros(1,1);

for j=1:m
    error1(1,j)=double(int((func1-bpfcoeff1(1,j))^2,
    ((j-1)*h),(j*h)));
    miseerrorbpf=miseerrorbpf+error1(1,j);
end
miseerrorbpf          % MISE for the coefficients
                      % obtained via direct expansion

miseerrorbpf_int=zeros(1,1);

for j=1:m
    error2(1,j)=double(int((func1-bpfcoeff_int(1,j))^2,
    ((j-1)*h),(j*h)));
    miseerrorbpf_int=miseerrorbpf_int+error2(1,j);
end

miseerrorbpf_int      % MISE for the coefficients obtained via
                      % operational matrix for integration
```

B.7 Program for Differentiation of a Function $f_1(t) = \sin(\pi t)$ Using Walsh/BPF Function for $m = 4$ (Example 3.2)

```
clc
clear all

%%- - - - - - - - - - - EXACT SOLUTION - - - - - - - - - - -%%

syms t
k=0.01;

FUNC1=diff(sin(pi*t))      % diferentiation of f(t)=sin(pi*t)

for i=1:101
    t1(i)=(i-1)*k;
    FUNC2(i)=pi*cos(pi*(i-1)*k);
end

        %%- - - - - - FUNCTION PLOTTING- - - - - - %%
```

```
plot(t1,FUNC2,'k')            % plot the exact function

%%- - - - - - - SOLUTION VIA DIRECT EXPANSION - - - - - - - -%%

syms t

func=sin(pi*t)       % given function
T=1                  % total time interval
m=4                  % no. of Walsh components
h=T/m                % length of each interval

W=[1  1   1   1;
   1  1  -1  -1;
   1 -1   1  -1;
   1 -1  -1   1];     % Walsh matrix

func1=diff(func)     % differentiation of the given function

for i=1:m
    x1(i)=double(m*int(func1,((i-1)*h),(i*h)));
                            % BPF coefficients of differentiated
                            % function via direct expansion
    bpfcoeff1(1,i)=x1(i);
end

Walshcoeff1=bpfcoeff1*inv(W)      % Walsh coefficents of the
                                  % differentiated
                            % function via direct expansion

%%- - SOLUTION VIA OPERATIONAL MATRIX FOR DIFFERENTIATION - -%%

opnmatrix_int=[1/2 -1/4 -1/8    0;
               1/4    0    0 -1/8;
               1/8    0    0    0;
                 0  1/8    0    0];
       % operational matrix for integration in Walsh domain
opnmatrix_diff=inv(opnmatrix_int);       % operational matrix for
                                         % differentiation in
                                         % Walsh domain
for i=1:m
    x1(i)=double(m*int(func,((i-1)*h),(i*h)));
    bpfcoeff(1,i)=x1(i);    %BPF coefficients of given function
end

Walshcoeff=bpfcoeff*inv(W);       % Walsh coefficents of the
                                  % given function

Walshcoeff_diff=Walshcoeff*opnmatrix_diff
             % Walsh coefficents of the
             % differentiated function via operational
             % matrix for differentiation in Walsh domain
```

```
bpfcoeff_diff=Walshcoeff_diff*W; % BPF coefficents of the
                                 % differentiated function
             % via operational matrix for differentiation in
             % Walsh domain

       %%- - - - - - FUNCTION PLOTTING- - - - - - %%

t=0:h:T;
hold on
stairs(t,[bpfcoeff1 0],'b')
hold on
stairs(t,[bpfcoeff_diff 0],'r')
hold on

%%- - - - PERCENTAGE ERROR BETWEEN WALSH COEFFICIENTS OBTAINED
%- VIA DIRECT EXPANSION AND OPERATIONAL MATRIX FOR
DIFFERENTIATION -%%

for i=1:m
percentageerror(i)=((Walshcoeff1(i)-Walshcoeff_diff(i))*100)/
                 Walshcoeff1(i);
end

percentageerror  % percentage error between Walsh coefficients
                 % obtained via direct expansion and operational
                 % matrix for integration

       %%- - - - - - - MISE CALCULATION- - - - - - -%%

miseerrorbpf=zeros(1,1);

for j=1:m
    error1(1,j)=double(int((func1-bpfcoeff1(1,j))^2,
    ((j-1)*h),(j*h)));
    miseerrorbpf=miseerrorbpf+error1(1,j) ;
end

miseerrorbpf         % MISE for the coefficients obtained via
                     % direct expansion

miseerrorbpf_diff=zeros(1,1);

for j=1:m
    error2(1,j)=double(int((func1-bpfcoeff_diff(1,j))^2,
    ((j-1)*h),(j*h)));
    miseerrorbpf_diff=miseerrorbpf_diff+error2(1,j) ;
end

miseerrorbpf_diff    % MISE for the coefficients obtained via
                     % operational matrix for integration
```

B.8 Program for Differentiation of a Function $f_1(t) = \sin(\pi t)$ Using Walsh/BPF Function for $m = 8$ (Example 3.2)

```
clc
clear all

%%- - - - - - - - - - - EXACT SOLUTION - - - - - - - - - - -%%

syms t
k=0.01;
FUNC1=diff(sin(pi*t))        % differentiation of f(t)=sin(pi*t)

for i=1:101
    t1(i)=(i-1)*k;
    FUNC2(i)=pi*cos(pi*(i-1)*k);
end

             %%- - - - - - FUNCTION PLOTTING- - - - - - %%

plot(t1,FUNC2,'k')           % plot the exact function

%%- - - - - - - SOLUTION VIA DIRECT EXPANSION - - - - - - - -%%

syms t
func=sin(pi*t)               % given function
T=1                          % total time considered
m=8                          % no. of Walsh components
h=T/m                        % length of each interval

W=[1  1  1  1  1  1  1  1;
   1  1  1  1 -1 -1 -1 -1;
   1  1 -1 -1  1  1 -1 -1;
   1  1 -1 -1 -1 -1  1  1;
   1 -1  1 -1  1 -1  1 -1;
   1 -1  1 -1 -1  1 -1  1;
   1 -1 -1  1  1 -1 -1  1;
   1 -1 -1  1 -1  1  1 -1] % Walsh matrix

func1=diff(func)     % differentiation of the given function

for i=1:m
    x1(i)=double(m*int(func1,((i-1)*h),(i*h)));
                    % BPF coefficients of differentiated
                    % function via direct expansion
    bpfcoeff1(1,i)=x1(i);
end

Walshcoeff1=bpfcoeff1*inv(W)      % Walsh coefficents of the
                                  % differentiated
                            % function via direct expansion
```

```
%%-- - SOLUTION VIA OPERATIONAL MATRIX FOR DIFFERENTIATION - -%%

opnmatrix_int=[1/2 -1/4 -1/8     0 -1/16     0     0     0;
                1/4    0    0 -1/8     0 -1/16     0     0;
                1/8    0    0    0     0     0 -1/16     0;
                  0  1/8    0    0     0     0     0 -1/16;
               1/16    0    0    0     0     0     0     0;
                  0 1/16    0    0     0     0     0     0;
                  0    0 1/16    0     0     0     0     0;
                  0    0    0 1/16    0     0     0     0]
```

                                   % operational matrix for
                                   % integration in Walsh domain

```
opnmatrix_diff=inv(opnmatrix_int);     % operational matrix for
                                       % differentiation in Walsh domain

for i=1:m
    x1(i)=double(m*int(func,((i-1)*h),(i*h)));
    bpfcoeff(1,i)=x1(i);   % BPF coefficients of given function
end
Walshcoeff=bpfcoeff*inv(W);              % Walsh coefficents of the
                                         % given function

Walshcoeff_diff=Walshcoeff*opnmatrix_diff
                % Walsh coefficents of the
                % differentiated function via operational
                % matrix for differentiation in Walsh domain

bpfcoeff_diff=Walshcoeff_diff*W;
                        % BPF coefficents of the differentiated
                        % function via operational matrix for
                        % differentiation in Walsh domain

        %%- - - - - - FUNCTION PLOTTING - - - - - - %%

t=0:h:T;
hold on
stairs(t,[bpfcoeff1 0],'b')
hold on
stairs(t,[bpfcoeff_diff 0],'r')
hold on

%%- - - - PERCENTAGE ERROR BETWEEN WALSH COEFFICIENTS OBTAINED
% VIA DIRECT EXPANSION AND OPERATIONAL MATRIX FOR
DIFFERENTIATION- - - %%

for i=1:m
    percentageerror(i)=((Walshcoeff1(i)-Walshcoeff_diff(i))*100)/
    Walshcoeff1(i);
end
```

```
percentageerror        % percentage error between Walsh
                       % coefficients
                       % obtained via direct expansion and
                       % operational
                       % matrix for integration

       %%- - - - - - - MISE CALCULATION- - - - - - -%%

miseerrorbpf=zeros(1,1);

for j=1:m
    error1(1,j)=double(int((func1-bpfcoeff1(1,j))^2,((j-1)*h),
    (j*h)));
    miseerrorbpf=miseerrorbpf+error1(1,j);
end

miseerrorbpf           % MISE for the coefficients obtained via
                       % direct expansion

miseerrorbpf_diff=zeros(1,1);

for j=1:m
    error2(1,j)=double(int((func1-bpfcoeff_diff(1,j))^2,((j-
    1)*h),(j*h)));
    miseerrorbpf_diff=miseerrorbpf_diff+error2(1,j);
end

miseerrorbpf_diff    % MISE for the coefficients obtained via
                     % operational matrix for integration
```

B.9 Analysis of First-Order System with Step Input for *m* = 8 (Section 3.4)

```
clc
clear all
format long

%%- - - - - - - - - - EXACT SOLUTION - - - - - - - - - - -%%

T=4                  % total time considered
m=8                  % no. of Walsh components considered
h=T/m                % length of each interval

for i=1:401
    func1(i)=1;      % func1=u(t)
end
```

```
for i=1:401
    t(i)=0.01*(i-1);
    func2(i)=exp(-t(i));    % func2=e^(-t)
end

func=func1-(func1*diag(func2));  % exact solution of the
                                 % function where,
                      % func=u(t)-u(t)exp(-t)

        %%- - - - - -|FUNCTION PLOTTING- - - - - - %%

for i=1:401
    t(i)=(i-1)*0.01;
end
plot(t,func,'r')               % plot the exact function

%%- - SOLUTION VIA WALSH FUNCTION USING WALSH OPERATIONAL
%%- - TRANSFER FUNCTION

lambda=T                       % scaling factor

W=[1  1  1  1  1  1  1  1;
   1  1  1  1 -1 -1 -1 -1;
   1  1 -1 -1  1  1 -1 -1;
   1  1 -1 -1 -1 -1  1  1;
   1 -1  1 -1  1 -1  1 -1;
   1 -1  1 -1 -1  1 -1  1;
   1 -1 -1  1  1 -1 -1  1;
   1 -1 -1  1 -1  1  1 -1]   % Walsh matrix

input_bpf=[1 1 1 1 1 1 1 1]        % BPF expansion of unit step
                                   % input for m=8

input_walsh=input_bpf*inv(W)       % Walsh expansion of unit
                                   % step input for m=8

%%- - - - - - - - - - - - - - - - - - - - - - - - - - -%%

opnint_walsh=[1/2 -1/4 -1/8    0 -1/16     0     0     0;
              1/4     0     0 -1/8     0 -1/16     0     0;
              1/8     0     0    0     0     0 -1/16     0;
                0   1/8     0    0     0     0     0 -1/16;
             1/16     0     0    0     0     0     0     0;
                0  1/16     0    0     0     0     0     0;
                0     0  1/16    0     0     0     0     0;
                0     0     0 1/16     0     0     0     0]
            % operational matrix for integration in Walsh domain

opndiff_walsh=inv(opnint_walsh)% operational matrix for
                               % differentiation in Walsh domain
```

```
WOTF1=inv(((1/lambda)*opndiff_walsh)+eye(m,m))  % Walsh
                                   % operational transfer
                        % function for first order system

output_walsh=input_walsh*WOTF1             % Walsh expansion of
                                           % the output of given
                                  % first order system

output_bpf=input_walsh*WOTF1*W              % BPF expansion of the
                                           % output of given
                                  % first order system

        %%- - - - - - FUNCTION PLOTTING- - - - - - %%

t=0:h:T
hold on
stairs(t,[output_bpf 0])              % plot the system output
```

B.10 Analysis of Second-Order System with Step Input for *m* = 8 (Section 3.5)

```
clc
clear all
format long

%%- - - - - - - - - -EXACT SOLUTION - - - - - - - - - - - -%%

T=8                    % total time considered
m=8                    % no. of Walsh components
h=T/m                  % length of each interval
wn=1                   % natural frequency of the system
z=0.3                  % damping ratio

for i=1:801
    t(i)=0.01*(i-1);
    func(i)=1-(exp(-(z*wn*t(i)))/sqrt(1-z^2))*sin(wn*(sqrt(1-
    z^2)*t(i))+atan(sqrt(1-z^2)/z))
                        % exact solution
end

        %%- - - - - - FUNCTION PLOTTING- - - - - - %%

for i=1:801
    t(i)=(i-1)*0.01;
end
plot(t,func,'r')              % plot the exact function

%%- - SOLUTION VIA WALSH FUNCTION USING WALSH OPERATIONAL
%%- - TRANSFER FUNCTION
```

```
lambda=T                          % scaling constant

W=[1   1   1   1   1   1   1   1;
   1   1   1   1  -1  -1  -1  -1;
   1   1  -1  -1   1   1  -1  -1;
   1   1  -1  -1  -1  -1   1   1;
   1  -1   1  -1   1  -1   1  -1;
   1  -1   1  -1  -1   1  -1   1;
   1  -1  -1   1   1  -1  -1   1;
   1  -1  -1   1  -1   1   1  -1]  % Walsh matrix

input_bpf=[1 1 1 1 1 1 1 1]        % BPF expansion of unit step
                                   % input for m=8

input_walsh=input_bpf*inv(W)       % Walsh expansion of unit
                                   % step input for m=8

%%- - - - - - - - - - - - --- - - - - - - - - - - - - - -%%

opnint_walsh=[1/2 -1/4 -1/8    0 -1/16     0      0      0;
             1/4     0    0 -1/8     0 -1/16      0      0;
             1/8     0    0    0     0     0 -1/16      0;
               0   1/8    0    0     0     0      0 -1/16;
            1/16     0    0    0     0     0      0      0;
               0  1/16    0    0     0     0      0      0;
               0     0 1/16    0     0     0      0      0;
               0     0    0 1/16     0     0      0      0]
        % operational matrix for integration in Walsh domain

opndiff_walsh=inv(opnint_walsh)% operational matrix for
                               % differentiation in Walsh domain

WOTF2=inv((opndiff_walsh^2/lambda^2)+((0.6*opndiff_walsh)/
lambda)+eye(m,m))
                    % Walsh operational transfer function for
                    % second order system

output_walsh=input_walsh*WOTF2     % Walsh expansion of the
                                   % output of given
                                   % second order system

output_bpf=input_walsh*WOTF2*W        % BPF expansion of the
                                      % output of given
                                      % second order system

        %%- - - - - - FUNCTION PLOTTING - - - - - - %%

t=0:h:T
hold on
stairs(t,[output_bpf 0])           % plot the system output
```

B.11 Oscillatory Phenomenon in First-Order System (Section 3.6.1)

```
clc
clear all

%%- - - - - - - - - - EXACT SOLUTION- - - - - - - - - - - - -%%

t=0.01
a=6
for i=1:401
    func(i)=(a^(-1))*(1-(exp(-a*(i-1)*t))); % equation (3.63)
end


        %%- - - - - - FUNCTION PLOTTING- - - - - - %%

for i=1:401
    t(i)=(i-1)*0.01;
end
hold on
plot(t,func,'r')            % plot the exact solution

%%- - - - - - - - SOLUTION VIA WALSH FUNCTION - - - - - - - -%%

clc
clear all
syms t
m=8                         % no. of Walsh components
lambda=4                    % scaling constant
a=6


        %%- - - - - - - program to find out D'

for i=1:m
    for j=1:m
        if i==j
            Dstar(i,j)=4*m*(1/2);
    elseif i>j
            Dstar(i,j)=0;
    else
        Dstar(i,j)=4*m*(-1)^(i+j);

    end
end
end
Dstar                       % equation (3.66)
```

```
%%- - - - - - - - - - - -

W=[1   1   1   1   1   1   1   1;
   1   1   1   1  -1  -1  -1  -1;
   1   1  -1  -1   1   1  -1  -1;
   1   1  -1  -1  -1  -1   1   1;
   1  -1   1  -1   1  -1   1  -1;
   1  -1   1  -1  -1   1  -1   1;
   1  -1  -1   1   1  -1  -1   1;
   1  -1  -1   1  -1   1   1  -1]                    % Walsh matrix
opndiff_walsh=(1/m)*W*Dstar*W

WOTF=inv(((1/lambda)*opndiff_walsh)+(a*eye(m,m)))  %Walsh
                              % operational transfer
                              % function for the given system

input_bpf=[1 1 1 1 1 1 1 1]    % BPF expansion of unit step
                              % input for m=8

input_walsh=input_bpf*inv(W)   % Walsh expansion of unit
                              % step input for m=8

output_walsh=input_walsh*(WOTF) % Walsh expansion of the
                              % output of given system

output_bpf=output_walsh*W       % BPF expansion of the
                              % output of given system

        %%- - - - - - FUNCTION PLOTTING- - - - - - %%

t=0:4/8:4;
stairs(t,[output_bpf 0])    % plot the solution via Walsh
                              % operational matrix
```

B.12 Program for Variation of Percentage Maximum Overshoot with Different Values of *a* Considering Different *m* and *λ* (Figures 3.12 through 3.16)

```
clc
clear all
syms t

%- - PERCENTAGE MAXIMUM OVERSHOOT FOR m=4 AND lambda=4 - -%

m=4                              % no. of Walsh components
lambda=4                         % scaling constant
```

```
for i=1:11
    a(i)=2+(i-1)*0.01;                 % different values of a
    p(i)=(a(i)*lambda)/(2*m);          % different values of p
overshoot(i)=((p(i)-1)/((p(i)+1)^2))*100;       % percentage max.
                                                % overshoot for
                                       % different values of a
end
plot(a,overshoot,'o-')  % plot a vs. percentage max. overshoot

%- - -PERCENTAGE MAXIMUM OVERSHOOT FOR m=8 AND lambda=4 - -%

m=8                               % no. of Walsh components
lambda=4                          % scaling constant

for i=1:25
    a(i)=4+(i-1)*0.1;                  % different values of a
    p(i)=(a(i)*lambda)/(2*m);          % different values of p
    overshoot(i)=((p(i)-1)/((p(i)+1)^2))*100; % percentage max.
                                       % overshoot for
                                       % different values of a
end
plot(a,overshoot,'o-')   % plot a vs. percentage max. overshoot

%- - - - - -PERCENTAGE MAXIMUM OVERSHOOT FOR m=16 AND lambda
=4- - - - - - %

m=16                               % no. of Walsh components
lambda=4                           % scaling constant

for i=1:16
    a(i)=8.00+(i-1)*0.2; % different values of a (choose different
                         % ranges of 'a' to observe the variation
                         % of overshoot)
    p(i)=(a(i)*lambda)/(2*m);          % different values of p
    overshoot(i)=((p(i)-1)/((p(i)+1)^2))*100;% percentage max.
                                       % overshoot for
                                       % different values of a
end
plot(a,overshoot,'o-')   % plot a vs. percentage max. overshoot
```

B.13 Analysis of a First-Order System with Single-Pulse Input for *m*=8 (Section 4.1.1)

```
clc
clear all
format short
```

```
%%- - - - - - - - - EXACT SOLUTION- - - - - - - - - - - -%%

T=4               % total time considered
m=8               % no. of Walsh components
h=T/m             % length of each interval

for i=1:401
    func1(i)=1                    %%% func1=u(t)
end

for i=1:401
    t(i)=(i-1)*0.01;
    if t<(h)
        func3(i)=0   %%% func3=u(t-h)
    else
        func3(i)=1
    end
end

for i=1:401
    t(i)=0.01*(i-1)
    func2(i)=exp(-t(i)); %%% func2=e^(-t)
    func4(i)=exp(h-(t(i))); %%% func4=e^(-t+h)
end

func8=func1-(func1*diag(func2))-(func3)+(func3*diag(func4))
      %%% func=u(t)-u(t)exp(-t)-u(t-h)+u(t-h)exp(-t+h)

        %%- - - - - - FUNCTION PLOTTING- - - - - - %%

for i=1:401
    t(i)=(i-1)*0.01;
end
subplot(2,1,2)
plot(t,func8,'r')

%%- - - - - - SOLUTION VIA WALSH FUNCTION
        % USING WALSH OPERATIONAL TRANSFER FUNCTION- - - %%

T=4                             % total time considered
m=8                             % no. of Walsh components
h=T/m                           % length of each interval
lambda=T                        % scaling factor

W=[1   1   1   1   1   1   1   1;
   1   1   1   1  -1  -1  -1  -1;
   1   1  -1  -1   1   1  -1  -1;
   1   1  -1  -1  -1  -1   1   1;
   1  -1   1  -1   1  -1   1  -1;
   1  -1   1  -1  -1   1  -1   1;
   1  -1  -1   1   1  -1  -1   1;
   1  -1  -1   1  -1   1   1  -1]        % Walsh matrix
```

```
input_bpf=[1 0 0 0 0 0 0 0]          % system input in BPF domain

input_walsh=input_bpf*inv(W)         % system input in Walsh domain

%%- - - - - - - - - - - - - - - - - - - - - - - - - - - - - -%%

opnint_walsh=[1/2 -1/4 -1/8     0 -1/16     0     0     0;
              1/4    0    0  -1/8     0 -1/16     0     0;
              1/8    0    0     0     0     0 -1/16     0;
                0  1/8    0     0     0     0     0 -1/16;
             1/16    0    0     0     0     0     0     0;
                0 1/16    0     0     0     0     0     0;
                0    0 1/16     0     0     0     0     0;
                0    0    0  1/16     0     0     0     0]
        % operational matrix for integration in Walsh domain

opndiff_walsh=inv(opnint_walsh)         % operational matrix for
                                        % differentiation in Walsh domain

WOTF1=inv(((1/lambda)*opndiff_walsh)+eye(m,m))  % Walsh
                            % operational transfer function

output_walsh=input_walsh*WOTF1              % system output in
                                            % Walsh domain
output_bpf=output_walsh*W

        %%- - - - - - - FUNCTION PLOTTING- - - - - - - - -%%

t=0:h:T
hold on
subplot(2,1,1)
stairs(t,[input_bpf 0])                  % plot system input
hold on
subplot(2,1,2)
stairs(t,[output_bpf 0])                 % plot system output
```

B.14 Analysis of a First-Order System with Pulse-Pair Input for $m = 8$ (Section 4.1.2)

```
clc
clear all
format short

%%- - - - - - - - - EXACT SOLUTION- - - - - - - - - - - -%%

T=4                 % total time considered
m=8                 % no. of Walsh components
```

```
h=T/m                   % length of each interval
lambda=T                % scaling factor

for i=1:401
    func1(i)=1      %%% func1=u(t)
end

for i=1:401
    t(i)=(i-1)*0.01;
    if t<(h)
        func3(i)=0    %%% func3=u(t-h)
    else
        func3(i)=1
    end
    if t<(4*h)
        func5(i)=0    %%% func5=u(t-4h)
    else
        func5(i)=1
    end
    if t<(5*h)
        func7(i)=0    %%% func7=u(t-5h)
    else
        func7(i)=1
    end
end

for i=1:401
    t(i)=0.01*(i-1)
    func2(i)=exp(-t(i));          %%% func2=e^(-t)
    func4(i)=exp(h-(t(i)));       %%% func4=e^(-t+h)
    func6(i)=exp((4*h)-(t(i)));   %%% func6=e^(-t+4h)
    func8(i)=exp((5*h)-(t(i)));   %%% func8=e^(-t+5h)
end

func=func1-(func1*diag(func2))-(func3)+(func3*diag(func4))
+func5-(func5*diag(func6))-(func7)+(func7*diag(func8))
%%% func=u(t)-u(t)exp(-t)-u(t-h)+u(t-h)exp(-t+h)+u(t-4h)
-u(t-4h)exp(-t+4h)-u(t-5h)+u(t-5h)exp(-t+5h)

        %%- - - - - - FUNCTION PLOTTING- - - - - - %%

for i=1:401
    t(i)=(i-1)*0.01;
end
subplot(2,1,2)
plot(t,func,'r')
hold on
```

```
%%- - - - - - SOLUTION VIA WALSH FUNCTION
      % USING WALSH OPERATIONAL TRANSFER FUNCTION- - - %%

T=4                                    % total time considered
m=8                                    % no. of Walsh components
h=T/m                                  % length of each interval
lambda=T                               % scaling factor

W=[1   1   1   1   1   1   1   1;
   1   1   1   1  -1  -1  -1  -1;
   1   1  -1  -1   1   1  -1  -1;
   1   1  -1  -1  -1  -1   1   1;
   1  -1   1  -1   1  -1   1  -1;
   1  -1   1  -1  -1   1  -1   1;
   1  -1  -1   1   1  -1  -1   1;
   1  -1  -1   1  -1   1   1  -1]       % Walsh matrix

input_bpf=[1 0 0 0 1 0 0 0]            % system input in BPF domain

input_walsh=input_bpf*inv(W)           % system input in Walsh domain

%%- - - - - - - - - - - - - - - - - - - - - - - - - - - - -%%

opnint_walsh=[1/2 -1/4 -1/8     0 -1/16     0     0      0;
              1/4     0     0 -1/8     0 -1/16     0      0;
              1/8     0     0    0     0     0 -1/16      0;
                0   1/8     0    0     0     0     0  -1/16;
             1/16     0     0    0     0     0     0      0;
                0  1/16     0    0     0     0     0      0;
                0     0  1/16    0     0     0     0      0;
                0     0     0 1/16     0     0     0      0]

        % operational matrix for integration in Walsh domain
opndiff_walsh=inv(opnint_walsh)
                          % operational matrix for
                          % differentiation in Walsh domain

WOTF1=inv(((1/lambda)*opndiff_walsh)+eye(8,8))  % Walsh
                          % operational transfer function

output_walsh=input_walsh*WOTF1             % system output in
                                           % Walsh domain

output_bpf=output_walsh*W

        %%- - - - - - - FUNCTION PLOTTING- - - - - - - -%%

t=0:h:T
hold on
```

```
subplot(2,1,1)
stairs(t,[input_bpf 0])              % plot system input
hold on
subplot(2,1,2)
stairs(t,[output_bpf 0])             % plot system output
```

B.15 Analysis of a First-Order System with Alternating Double-Pulse Input for *m* = 8 (Section 4.1.3)

```
clc
clear all
format short

%%- - - - - - - - - - EXACT SOLUTION- - - - - - - - - - - -%%

T=4                   % total time considered
m=8                   % no. of Walsh components
h=T/m                 % length of each interval
lambda=T              % scaling factor

for i=1:401
    func1(i)=1              %%% func1=u(t)
end

for i=1:401
    t(i)=(i-1)*0.01;
    if t<(h)
        func3(i)=0         %%% func3=u(t-h)
    else
        func3(i)=1
    end
    if t<(2*h)
        func5(i)=0         %%% func5=u(t-2h)
    else
        func5(i)=1
    end
    if t<(3*h)
        func7(i)=0         %%% func7=u(t-3h)
    else
        func7(i)=1
    end
    if t<(4*h)
        func9(i)=0         %%% func9=u(t-4h)
    else
        func9(i)=1
    end
    if t<(5*h)
        func11(i)=0        %%% func11=u(t-5h)
```

```
      else
         func11(i)=1
      end
end


for i=1:401
    t(i)=0.01*(i-1)
    func2(i)=exp(-t(i));                %%% func2=e^(-t)
    func4(i)=exp(h-(t(i)));             %%% func4=e^(-t+h)
    func6(i)=exp((2*h)-(t(i)));         %%% func6=e^(-t+2h)
    func8(i)=exp((3*h)-(t(i)));         %%% func8=e^(-t+3h)
    func10(i)=exp((4*h)-(t(i)));        %%% func10=e^(-t+4h)
    func12(i)=exp((5*h)-(t(i)));        %%% func12=e^(-t+5h)
end


func=func1-(func1*diag(func2))-(2*func3)+(2*func3*diag(func4))
+func5-(func5*diag(func6))+(func7)-(func7*diag(func8))-(2*func
9)+(2*func9*diag(func10))+func11-(func11*diag(func12))
%%% func=u(t)-u(t)exp(-t)-2u(t-h)+2u(t-h)exp(-t+h)+u(t-2h)
-u(t-2h)exp(-t+2h)+u(t-3h)-u(t-3h)exp(-t+3h)-2u(t-4h)+2u(t-4h)
exp(-t+4h)+u(t-5h)-u(t-5h)exp(-t+5h)


        %%- - - - - - FUNCTION PLOTTING- - - - - - %%


for i=1:401
    t(i)=(i-1)*0.01;
end
subplot(2,1,2)
plot(t,func,'r')
hold on


%%- - - - - - SOLUTION VIA WALSH FUNCTION
        % USING WALSH OPERATIONAL TRANSFER FUNCTION- - - %%


T=4                                 % total time considered
m=8
h=T/m                               % length of each interval
lambda=T                            % scaling factor


W=[1  1  1  1  1  1  1  1;
   1  1  1  1 -1 -1 -1 -1;
   1  1 -1 -1  1  1 -1 -1;
   1  1 -1 -1 -1 -1  1  1;
   1 -1  1 -1  1 -1  1 -1;
   1 -1  1 -1 -1  1 -1  1;
   1 -1 -1  1  1 -1 -1  1;
   1 -1 -1  1 -1  1  1 -1]          % Walsh matrix
```

```
input_bpf=[1 -1 0 1 -1 0 0 0]     % system input in BPF domain

input_walsh=input_bpf*inv(W)      % system input in Walsh domain

%%— — — — — — — — — — — — — — — — — — — — — — — — — — -%%

opnint_walsh=[1/2 -1/4 -1/8    0 -1/16    0      0      0;
              1/4    0    0 -1/8     0 -1/16     0      0;
              1/8    0    0    0     0     0 -1/16     0;
                0  1/8    0    0     0     0     0 -1/16;
             1/16    0    0    0     0     0     0      0;
                0 1/16    0    0     0     0     0      0;
                0    0 1/16    0     0     0     0      0;
                0    0    0 1/16     0     0     0      0]

       % operational matrix for integration in Walsh domain
opndiff_walsh=inv(opnint_walsh)       % operational matrix for
                           % differentiation in Walsh domain

WOTF1=inv(((1/lambda)*opndiff_walsh)+eye(8,8)) % Walsh
                           % operational transfer function

output_walsh=input_walsh*WOTF1            % system output in
                                          % Walsh domain

output_bpf=output_walsh*W

       %%— — — — — — - FUNCTION PLOTTING— — — — — — — — -%%

t=0:h:T
hold on
subplot(2,1,1)
stairs(t,[input_bpf 0])           % plot system input
hold on
subplot(2,1,2)
stairs(t,[output_bpf 0])          % plot system output
```

B.16 Analysis of a Second-Order System with Single-Pulse Input for $m = 8$ (Section 4.2.1)

```
clc
clear all
format short

%%— — — — — — — — — — EXACT SOLUTION— — — — — — — — — — -%%

T=8                    % total time considered
m=8                    % no. of Walsh components
h=T/m                  % length of each interval
```

```
wn=1                    % natural frequency
z=0.3                   % damping coefficient

for i=1:401
    t(i)=0.02*(i-1);
    if t<(h)
        func2(i)=0    %%% func3=u(t-h)
    else
        func2(i)=1
    end
end

for i=1:401
    t(i)=0.02*(i-1);
    func3(i)=exp(z*wn*(h-(t(i))))*cos(wn*(sqrt(1-
    z^2))*(t(i)-h)); %%% func4=e^(-t+h)
    func4(i)=((z*wn)/(wn*(sqrt(1-z^2))))*exp(z*wn*(h-
    (t(i))))*sin(wn*(sqrt(1-z^2))*(t(i)-h));   %%% func6=e^(-
    t+4h)
end
func5=func2-(func2*diag(func3))-(func2*diag(func4))

for i=1:401
    t(i)=0.02*(i-1);
    func1(i)=1-(exp(-(z*wn*t(i)))/sqrt(1-z^2))*sin(wn*(sqrt(1-
    z^2)*t(i))+atan(sqrt(1-z^2)/z))
end

func=func1-func5              % exact solution

        %%- - - - - - FUNCTION PLOTTING- - - - - - %%

for i=1:401
    t(i)=(i-1)*0.02;
end
subplot(2,1,2)
plot(t,func,'r')             % plot the exact function

%%- - - - - - SOLUTION VIA WALSH FUNCTION
        % USING WALSH OPERATIONAL TRANSFER FUNCTION- - - %%

lambda=T                     % scaling factor

W=[1  1  1  1  1  1  1  1;
   1  1  1  1 -1 -1 -1 -1;
   1  1 -1 -1  1  1 -1 -1;
   1  1 -1 -1 -1 -1  1  1;
   1 -1  1 -1  1 -1  1 -1;
   1 -1  1 -1 -1  1 -1  1;
   1 -1 -1  1  1 -1 -1  1;
   1 -1 -1  1 -1  1  1 -1]     % Walsh matrix
```

```
input_bpf=[1 0 0 0 0 0 0 0]          % BPF expansion of unit step
                                     % input for m=8


input_walsh=input_bpf*inv(W)         % Walsh expansion of unit
                                     % step input for m=8


%%- - - - - - - - - - - - - - - - - - - - - - - - - - - - - -%%


opnint_walsh=[1/2 -1/4 -1/8    0 -1/16     0      0     0;
              1/4    0    0 -1/8     0 -1/16      0     0;
              1/8    0    0    0     0      0 -1/16     0;
                0  1/8    0    0     0      0     0 -1/16;
             1/16    0    0    0     0      0     0     0;
                0 1/16    0    0     0      0     0     0;
                0    0 1/16    0     0      0     0     0;
                0    0    0 1/16     0      0     0     0]

       % operational matrix for integration in Walsh domain


opndiff_walsh=inv(opnint_walsh)      % operational matrix for
                                     % differentiation
                                     % in Walsh domain


WOTF2=inv((opndiff_walsh^2/lambda^2)+((0.6*opndiff_walsh)/
lambda)+eye(8,8))
                    % Walsh operational transfer function for second
                    % order system


output_walsh=input_walsh*WOTF2       % Walsh expansion of the
                                     % output of given
                                     % second order system


output_bpf=input_walsh*WOTF2*W             % BPF expansion of the
                                           % output of given
                                           % second order system


          %%- - - - - - - FUNCTION PLOTTING- - - - - - - -%%


t=0:h:T
hold on
subplot(2,1,1)
stairs(t,[input_bpf 0])                    % plot system input
hold on
subplot(2,1,2)
stairs(t,[output_bpf 0])                   % plot system output
```

B.17 Analysis of a Second-Order System with Pulse-Pair Input for $m = 8$ (Section 4.2.2)

```
clc
clear all
format short

%%- - - - - - - - - - EXACT SOLUTION - - - - - - - - - - - - -%%

T=8                      % total time considered
m=8                      % no. of Walsh components
h=T/m                    % length of each interval
wn=1                     % natural frequency
z=0.3                    % damping coefficient

for i=1:401
    t(i)=0.02*(i-1);
    if t<(h)
       func2(i)=0              %%% func3=u(t-h)
    else
       func2(i)=1
    end
    if t<(4*h)
       func6(i)=0              %%% func3=u(t-4h)
    else
       func6(i)=1
    end
    if t<(5*h)
       func10(i)=0             %%% func3=u(t-5h)
    else
       func10(i)=1
    end
end

for i=1:401
    t(i)=0.02*(i-1);
    func3(i)=exp(z*wn*(h-(t(i))))*cos(wn*(sqrt(1-z^2))
    *(t(i)-h));                          %%% func4=e^(-t+h)
    func4(i)=((z*wn)/(wn*(sqrt(1-z^2))))*exp(z*wn*(h-(t(i))))
    *sin(wn*(sqrt(1-z^2))*(t(i)-h));     %%% func6=e^(-t+4h)
end

func5=func2-(func2*diag(func3))-(func2*diag(func4));

for i=1:401
    t(i)=0.02*(i-1);
    func7(i)=exp(z*wn*(4*h-(t(i))))*cos(wn*(sqrt(1-z^2))
    *(t(i)-4*h));
                                         %%% func4=e^(-t+h)
```

```
        func8(i)=((z*wn)/(wn*(sqrt(1-z^2))))*exp(z*wn*(4*h-
        (t(i))))*sin(wn*(sqrt(1-z^2))*(t(i)-4*h));
                                        %%% func6=e^(-t+4h)
end

func9=func6-(func6*diag(func7))-(func6*diag(func8))

for i=1:401
    t(i)=0.02*(i-1);
    func11(i)=exp(z*wn*(5*h-(t(i))))*cos(wn*(sqrt(1-
    z^2))*(t(i)-5*h));
                                        %%% func4=e^(-t+h)

    func12(i)=((z*wn)/(wn*(sqrt(1-z^2))))*exp(z*wn*(5*h-
    (t(i))))*sin(wn*(sqrt(1-z^2))*(t(i)-5*h));
                                        %%% func6=e^(-t+4h)

end

func13=func10-(func10*diag(func11))-(func10*diag(func12))

for i=1:401
    t(i)=0.02*(i-1);
    func1(i)=1-(exp(-(z*wn*t(i)))/sqrt(1-z^2))*sin(wn*(sqrt(1-
    z^2)*t(i))+atan(sqrt(1-z^2)/z))
end

func=func1-func5+func9-func13            % exact solution

        %%- - - - - - FUNCTION PLOTTING- - - - - - %%

for i=1:401
    t(i)=(i-1)*0.02;
end
subplot(2,1,2)
plot(t,func,'r')                % plot the exact function

%%- - - - - - SOLUTION VIA WALSH FUNCTION
        % USING WALSH OPERATIONAL TRANSFER FUNCTION- - - %%

lambda=T                    % scaling factor

W=[1   1   1   1   1   1   1   1;
   1   1   1   1  -1  -1  -1  -1;
   1   1  -1  -1   1   1  -1  -1;
   1   1  -1  -1  -1  -1   1   1;
   1  -1   1  -1   1  -1   1  -1;
   1  -1   1  -1  -1   1  -1   1;
   1  -1  -1   1   1  -1  -1   1;
   1  -1  -1   1  -1   1   1  -1]        % Walsh matrix
```

```
input_bpf=[1 0 0 0 1 0 0 0]        % system input in BPF domain


input_walsh=input_bpf*inv(W)       % Walsh expansion of unit
                                   % step input for m=8


%%- - - - - - - - - - - - - - - - - - - - - - - - - - - - - - -%%


opnint_walsh=[1/2 -1/4 -1/8      0 -1/16      0      0    0;
              1/4    0    0   -1/8      0  -1/16      0    0;
              1/8    0    0      0      0      0  -1/16    0;
                0  1/8    0      0      0      0      0 -1/16;
             1/16    0    0      0      0      0      0    0;
                0 1/16    0      0      0      0      0    0;
                0    0 1/16      0      0      0      0    0;
                0    0    0   1/16      0      0      0    0]
               % operational matrix for integration in Walsh domain


opndiff_walsh=inv(opnint_walsh)    % operational matrix for
                                   % differentiation
                                   % in Walsh domain


WOTF2=inv((opndiff_walsh^2/lambda^2)+((0.6*opndiff_walsh)/
lambda)+eye(8,8))
                    % Walsh operational transfer function for
                    % second order system


output_walsh=input_walsh*WOTF2     % Walsh expansion of the
                                   % output of given
                                   % second order system


output_bpf=input_walsh*WOTF2*W              % BPF expansion of the
                                            % output of given
                                            % second order system


        %%- - - - - - - FUNCTION PLOTTING- - - - - - - - -%%


t=0:h:T
hold on
subplot(2,1,1)
stairs(t,[input_bpf 0])            % plot system input
hold on
subplot(2,1,2)
stairs(t,[output_bpf 0])           % plot system output
```

B.18 Analysis of a Second-Order System with Alternating Double-Pulse Input for *m* = 8 (Section 4.2.3)

```
clc
clear all
format short

%%- - - - - - - - - - EXACT SOLUTION- - - - - - - - - - - - -%%

T=8                   % total time considered
m=8                   % no. of Walsh components
h=T/m                 % length of each interval
wn=1                  % natural frequency
z=0.3                 % damping coefficient

for i=1:401
    t(i)=(i-1)*0.02;
    if t<(h)
        func2(i)=0    %%% func2=u(t-h)
    else
        func2(i)=1
    end
    if t<(2*h)
        func6(i)=0    %%% func6=u(t-2h)
    else
        func6(i)=1
    end
    if t<(3*h)
        func10(i)=0   %%% func10=u(t-3h)
    else
        func10(i)=1
    end
    if t<(4*h)
        func14(i)=0   %%% func14=u(t-4h)
    else
        func14(i)=1
    end
    if t<(5*h)
        func18(i)=0   %%% func18=u(t-5h)
    else
        func18(i)=1
    end
end

for i=1:401
    t(i)=0.02*(i-1);
    func3(i)=exp(z*wn*((h)-(t(i))))*cos(wn*(sqrt(1-z^2))
    *(t(i)-(h)));
```

```
        func4(i)=((z*wn)/(wn*(sqrt(1-z^2))))*exp(z*wn*((h)-
        (t(i))))*sin(wn*(sqrt(1-z^2))*(t(i)-(h)));
end


func5=2*(func2-(func2*diag(func3))-(func2*diag(func4)))


for i=1:401
        t(i)=0.02*(i-1);
        func7(i)=exp(z*wn*((2*h)-(t(i))))*cos(wn*(sqrt(1-
        z^2))*(t(i)-(2*h)));
        func8(i)=((z*wn)/(wn*(sqrt(1-z^2))))*exp(z*wn*((2*h)-
        (t(i))))*sin(wn*(sqrt(1-z^2))*(t(i)-(2*h)));
end


func9=func6-(func6*diag(func7))-(func6*diag(func8))


for i=1:401
        t(i)=0.02*(i-1);
        func11(i)=exp(z*wn*((3*h)-(t(i))))*cos(wn*(sqrt(1-
        z^2))*(t(i)-(3*h)));
        func12(i)=((z*wn)/(wn*(sqrt(1-z^2))))*exp(z*wn*((3*h)-
        (t(i))))*sin(wn*(sqrt(1-z^2))*(t(i)-(3*h)));
end


func13=func10-(func10*diag(func11))-(func10*diag(func12))


for i=1:401
        t(i)=0.02*(i-1);
        func15(i)=exp(z*wn*((4*h)-(t(i))))*cos(wn*(sqrt(1-
        z^2))*(t(i)-(4*h)));
        func16(i)=((z*wn)/(wn*(sqrt(1-z^2))))*exp(z*wn*((4*h)-
        (t(i))))*sin(wn*(sqrt(1-z^2))*(t(i)-(4*h)));
end


func17=2*(func14-(func14*diag(func15))-(func14*diag(func16)))


for i=1:401
        t(i)=0.02*(i-1);
        func19(i)=exp(z*wn*((5*h)-(t(i))))*cos(wn*(sqrt(1-
        z^2))*(t(i)-(5*h)));
        func20(i)=((z*wn)/(wn*(sqrt(1-z^2))))*exp(z*wn*((5*h)-
        (t(i))))*sin(wn*(sqrt(1-z^2))*(t(i)-(5*h)));
end


func21=func18-(func18*diag(func19))-(func18*diag(func20))


for i=1:401
        t(i)=0.02*(i-1);
```

```
    func1(i)=1-(exp(-(z*wn*t(i)))/sqrt(1-z^2))*sin(wn*(sqrt(1-
    z^2)*t(i))+atan(sqrt(1-z^2)/z))
end

func=func1-func5+func9+func13-func17+func21      % exact solution

    %%- - - - - - FUNCTION PLOTTING- - - - - - %%

for i=1:401
    t(i)=(i-1)*0.02;
end

subplot(2,1,2)
plot(t,func,'r')                % plot the exact function

%%- - - - - - SOLUTION VIA WALSH FUNCTION
        % USING WALSH OPERATIONAL TRANSFER FUNCTION- - - %%

lambda=T                        % scaling factor

W=[1   1   1   1   1   1   1   1;
   1   1   1   1  -1  -1  -1  -1;
   1   1  -1  -1   1   1  -1  -1;
   1   1  -1  -1  -1  -1   1   1;
   1  -1   1  -1   1  -1   1  -1;
   1  -1   1  -1  -1   1  -1   1;
   1  -1  -1   1   1  -1  -1   1;
   1  -1  -1   1  -1   1   1  -1]        % Walsh matrix

input_bpf=[1 -1 0 1 -1 0 0 0]           % BPF expansion of alternating
                                        % double pulse input for m=8

input_walsh=input_bpf*inv(W)            % Walsh expansion of unit
                                        % step input for m=8

%%- - - - - - - - - - - - - - - - - - - - - - - - - - -%%

opnint_walsh=[1/2 -1/4 -1/8     0 -1/16     0      0      0;
              1/4     0     0 -1/8     0 -1/16      0      0;
              1/8     0     0     0     0      0 -1/16      0;
                0   1/8     0     0     0      0      0 -1/16;
             1/16     0     0     0     0      0      0      0;
                0  1/16     0     0     0      0      0      0;
                0     0  1/16     0     0      0      0      0;
                0     0     0  1/16     0      0      0      0]

    % operational matrix for integration in Walsh domain
opndiff_walsh=inv(opnint_walsh)         % operational matrix for
                                        % differentiation in Walsh domain
```

```
WOTF2=inv((opndiff_walsh^2/lambda^2)+((0.6*opndiff_walsh)/
lambda)+eye(8,8))
                % Walsh operational transfer function for second
                             % order system

output_walsh=input_walsh*WOTF2     % Walsh expansion of the
                                   % output of given
                                   % second order system

output_bpf=input_walsh*WOTF2*W     % BPF expansion of the
                                   % output of given
                                   % second order system

        %%- - - - - - - FUNCTION PLOTTING- - - - - - - - -%%

t=0:h:T
hold on
subplot(2,1,1)
stairs(t,[input_bpf 0])            % plot system input
hold on
subplot(2,1,2)
stairs(t,[output_bpf 0])           % plot system output
```

B.19 Program to Determine Average and rms Values of Load Current, Device Current, and Diode Current for a Given Input Voltage to an R–L Load for $m = 8$ (Section 4.3.1.1, Figure 4.8, Tables 4.1 and 4.2)

```
clc
clear all
format short
V=1             % normalized input voltage
R=31            % load resistance
L=0.17          % load inductance
T=0.08          % Total time considered

m=8             % no. of segments
h=T/m           % length of each interval

%%- - - - - - - - - - - - - - - - - - - - - - - - - - - - -%%

P=[1  0  0  0  1  0  0  0;
   1  1  0  0  1  1  0  0;
   1  1  1  0  1  1  1  0]    % normalized input voltage in the
                      % time interval 0-T for KD=0.25,0.50
                      % and 0.75 respectively
PWMM=V*P              % pulse width modulation matrix
```

```
%%- - - - - - - - - - - - - - - - - - - - - - - - - - - - -%%

for i=1:m
    for j=1:m
        if i==j
            Dstar(i,j)=4*m*(1/2);
        elseif i>j
                Dstar(i,j)=0;
        else
            Dstar(i,j)=4*m*(-1)^(i+j);

        end
    end
end
Dstar

%%- - - - - - - - - - - - - - - - - - - - - - - - - - - -%%

W=[1  1  1  1  1  1  1  1;
   1  1  1  1 -1 -1 -1 -1;
   1  1 -1 -1  1  1 -1 -1;
   1  1 -1 -1 -1 -1  1  1;
   1 -1  1 -1  1 -1  1 -1;
   1 -1  1 -1 -1  1 -1  1;
   1 -1 -1  1  1 -1 -1  1;
   1 -1 -1  1 -1  1  1-1];                    % Walsh matrix

opndiff_walsh=(1/m)*W*Dstar*W    % operational matrix for
                                 % differentiation in Walsh domain

%%- - - - - - - - - - - - - - - - - - - - - - - - - - - -%%

WOTF=R*(inv((R*(eye(m,m)))+((L)*(opndiff_walsh/(T)))))
                    % Walsh operational transfer function

input_walsh=(1/m)*PWMM*W          % input in Walsh domain
output_walsh=input_walsh*WOTF     % output in Walsh domain
output_bpf=output_walsh*W         % output in BPF domain

%%- - PLOTTING LOAD CURRENT FOR KD=0.25, 0.50 AND 0.75 - -%%

t=0:h:T
for i=1:3
    for j=1:m
        output_bpf1(j)=[output_bpf(i,j)]
    end
    subplot(3,1,i)
    stairs(t,[output_bpf1 0],'k')% plotting the coefficients
                                 % obtained via
                                 % Walsh function
```

```
hold on
end

%%- - - - - - - - - - LOAD CURRENT- - - - - - - - - - - -%%

for i=1:3
    output_bpf1=0;
    output_bpf2=0;
    for j=1:m
        output_bpf1=[output_bpf(i,j)+output_bpf1];
        output_bpf2=[output_bpf(i,j)^2+output_bpf2];
    end
    avg_load_current(i)=(output_bpf1)/m;
    rms_load_current(i)=sqrt((output_bpf2)/m);
end

avg_load_current           % average load current obtained in
                           % Walsh domain
rms_load_current           % rms load current obtained in
                           % Walsh domain

%%- - - - - - - - - DEVICE CURRENT- - - - - - - - - - - -%%

for i=1:3
    for j=1:m
        output_bpf1(j)=[output_bpf(i,j)];
        output_bpf2(j)=[output_bpf(i,j)^2];
        P1(j)=[P(i,j)];
    end
    avg_device_current(i)=(output_bpf1*transpose(P1))/m;
    rms_device_current(i)=sqrt((output_bpf2*transpose(P1))/m)
end
avg_device_current  % average device current obtained in
                    % Walsh domain
rms_device_current  % rms device current obtained in
                    % Walsh domain

%%- - - - - - - - - - FREEWHEELING CURRENT- - - - - - - - -%%

for i=1:3
    for j=1:m
        output_bpf1(j)=[output_bpf(i,j)];
        output_bpf2(j)=[output_bpf(i,j)^2];
        P1(j)=1-[P(i,j)];
    end
    avg_diode_current(i)=(output_bpf1*transpose(P1))/m
    rms_diode_current(i)=sqrt((output_bpf2*transpose(P1))/m)
end

avg_diode_current          % average diode current obtained in
                           % Walsh domain
```

```
rms_diode_current      % rms diode current obtained in
                       % Walsh domain

%%- - - - EXACT SOLUTION OF AVERAGE AND RMS CURRENTS- - - - %%

Tr=0.04                          % repetition rate of input waveform
Tl=L/R                           % load time constant
a=Tr/Tl
Kd=0.25:0.25:3*0.25              % duty cycle of chopper KD=0.25,
                                 % 0.50 and 0.75
Kf=1-Kd                          % freewheeling cycle of chopper
for k=1:3
    avg_device_exact(k)=Kd(k)-((1-exp(-(a*Kd(k))))/a)
    +((1-exp(-(a*Kd(k))))^2*(exp(-(a*Kf(k)))/(2*a)));
                                            % equation (4.28)

    avg_diode_exact(k)=(1-exp(-(a*Kd(k))))*(1-exp(-
    (a*Kf(k))))*((2+exp(-a))/(2*a));        % equation (4.30)
    avg_load_exact=avg_device_exact+ avg_diode_exact;
                                            % equation (4.31)

    A(k)=(1-exp(-(a*Kd(k))))/a;

    B(k)=(1-exp(-(2*a*Kd(k))))/(4*a);

    rms_device_exact(k)=sqrt(Kd(k)-(2*A(k))+(2*B(k))-
    (2*A(k)*B(k)*a*exp(-(a*Kf(k))))+((a^2)*(A(k)^2)*B(k)*
    exp(-(2*a*Kf(k))))+(a*(A(k)^2)*exp(-(a*Kf(k)))));
                                            % equation (4.33)

    rms_diode_exact(k)=(1/2)*(sqrt((a*A(k)^2)*(1-exp(-
    (2*a*Kf(k))))*(2+(2*exp(-a))+exp(-(2*a)))));
                                            % equation (4.34)

    rms_load_exact(k)=sqrt(rms_device_exact(k)^2+rms_diode_
    exact(k)^2);
                                            % equation (4.35)
end

avg_load_exact               % average load current obtained
                             % via exact solution
avg_device_exact             % average device current obtained
                             % via exact solution
avg_diode_exact              % average diode current obtained
                             % via exact solution
rms_load_exact               % rms load current obtained
                             % via exact solution
rms_device_exact             % rms device current obtained
                             % via exact solution
rms_diode_exact              % rms diode current obtained
                             % via exact solution
```

```
%%- - — - PLOTTING LOAD CURRENT FOR Kd=0.25, 0.50 AND 0.75
             % VIA EXACT SOLUTION- - - - - - - - - - - -%%

for k=1:3
    for i=1:81
    t(i)=(i-1)*0.001;
    if t(i)<(Kd(k)*Tr)
        func8(i)=1;           %%% func8=u(t-Kd*Tr)
    else
        func8(i)=0;
    end

    if t(i)<(Kd(k)*Tr)
        func1(i)=0;           %%% func1=u(t-Kd*Tr)
    else
        func1(i)=1;
    end

    if t(i)<(Tr+(Kd(k)*Tr))
        func2(i)=0;           %%% func2=u(t-Tr-Kd*Tr)
    else
        func2(i)=1;
    end

    if t(i)<(Tr)
        func5(i)=0;           %%% func5=u(t-Tr)
    else
        func5(i)=1;
    end
end
func=func5-func2 ;           %%% func=u(t-Tr)-u(t-Tr-Kd*Tr)

for i=1:81
    t(i)=0.001*(i-1);
    func3(i)=exp(-(R*(t(i)-(Kd(k)*Tr)))/L);
    func4(i)=exp(-(R*(t(i)-Tr-(Kd(k)*Tr)))/L);
end

If1=(1-exp(-(a*Kd(k))))*func3*diag(func1);
                        % If1=((1-exp(-(a*Kd)))*exp(-(R*
                        % (t-(Kd*Tr))/L)))*u(t-(Kd*Tr))

If2=(1-exp(-(a*Kd(k))))*(exp(-a)+1)*func4*diag(func2);
%If2=((1-exp(-(a*Kd)))*(exp(-a)+1)*exp(-(R*(t-Tr-
                        % (Kd*Tr))/L)))*u(t-Tr-(Kd*Tr))

If=If1+If2;          % freewheeling current from equation (4.26)

for i=1:81
    t(i)=0.001*(i-1);
```

```
    func6(i)=(1-exp(-(R*t(i))/L));
    func7(i)=((1-exp(-(a*Kd(k))))*(exp(-(a*Kf(k))))*(exp(-
    (R*(t(i)-Tr))/L)))+(1-exp(-(R*(t(i)-Tr))/L));
end
    Is1=func6*diag(func8);  % Is1=((1-exp(-(R*t)/L))*u(t))
    Is2=func7*diag(func);   % Is2=((1-exp(-(a*Kd)))*(exp-
    (a*Kf))*(exp(-(R*(t-
                  % Tr))/L)))+(1-exp(-(R*(t-Tr))/L)))*u(t-Tr)

    Is=Is1+Is2;            % device current from equation (4.25)

    Il=If+Is;              % load current

    hold on
    subplot(3,1,k)
    plot(t,Il,'k')
    hold on
end
```

B.20 Program to Determine Average and rms Values of Load Current, Device Current, and Diode Current for a Given Input Voltage to an RL Load for *m* = 16 (Section 4.3.1.1, Figure 4.12, Tables 4.3 and 4.4)

```
clc
clear all
format short

V=1                     % normalized input voltage
R=31                    % load resistance
L=0.17                  % load inductance
T=0.08                  % Total time considered
m=16                    % no. of segments
h=T/m                   % length of each interval
Kd=transpose([0.1 0.2 0.25 0.3 0.4 0.5 0.6 0.7 0.75 0.8 0.9
1.0])                   % duty cycle considered
p=size(Kd)

%%- - - - - - - SOLUTION VIA WALSH FUNCTION- - - - - - - - %%

%%- - - - - PULSE WIDTH MODULATION MATRIX FORMATION- - - - -%%%

for i=1:p(1,1)
    Kd1=Kd(i,1)         % select one duty cycle for each iteration
f=(m/2)*Kd1;
```

```
n=fix(f);
x(i)=f-fix(f)         % fraction x at (n+1)th subinterval
if n==0
      P(1,n+1)=x(i);        % area preservation for (n+1)th
                            % subinterval
else
      n=n;
      P=ones(1,n);          % input for on time of the device
      if x(i)~=0
         P(1,n+1)=x(i);     % area preservation for (n+1)th
                            % subinterval
      end
end

q=size(P);
PP=zeros(1,(m/2)-q(1,2));  % input for the off time of the
                          % device

coeff1=[P PP];

coeff2=[coeff1 coeff1];   % input coefficient for individual
                          % duty cycle

for j=1:m

      PWMM1(i,j)=coeff2(1,j);   % input coefficients are
                                % assigned in pulse
                                % width modulation matrix
end

for j=1:3
    if Kd1==0.25*j                  % duty cycle Kd=0.25, 0.50
                                    % and 0.75 required
                                    % for function plotting
    d(j)=i
    end
end

end

PWMM=V*PWMM1                  % pulse width modulation matrix

%%— — — — — — — — — — — — — — — — — — — — — — — — — — — — — -%%

for i=1:m
    for j=1:m
        if i==j
                Dstar(i,j)=4*m*(1/2);
        elseif i>j
```

```
        Dstar(i,j)=0;
    else
        Dstar(i,j)=4*m*(-1)^(i+j);

    end
  end
end
Dstar;
```

%%— -%%

```
W=[1  1  1  1  1  1  1  1  1  1  1  1  1  1  1  1;
   1  1  1  1  1  1  1  1 -1 -1 -1 -1 -1 -1 -1 -1;
   1  1  1  1 -1 -1 -1 -1  1  1  1  1 -1 -1 -1 -1;
   1  1  1  1 -1 -1 -1 -1 -1 -1 -1 -1  1  1  1  1;
   1  1 -1 -1  1  1 -1 -1  1  1 -1 -1  1  1 -1 -1;
   1  1 -1 -1  1  1 -1 -1 -1 -1  1  1 -1 -1  1  1;
   1  1 -1 -1 -1 -1  1  1  1  1 -1 -1 -1 -1  1  1;
   1  1 -1 -1 -1 -1  1  1 -1 -1  1  1  1  1 -1 -1;
   1 -1  1 -1  1 -1  1 -1  1 -1  1 -1  1 -1  1 -1;
   1 -1  1 -1  1 -1  1 -1 -1  1 -1  1 -1  1 -1  1;
   1 -1  1 -1 -1  1 -1  1  1 -1  1 -1 -1  1 -1  1;
   1 -1  1 -1 -1  1 -1  1 -1  1 -1  1  1 -1  1 -1;
   1 -1 -1  1  1 -1 -1  1  1 -1 -1  1  1 -1 -1  1;
   1 -1 -1  1  1 -1 -1  1 -1  1  1 -1 -1  1  1 -1;
   1 -1 -1  1 -1  1  1 -1  1 -1 -1  1 -1  1  1 -1;
   1 -1 -1  1 -1  1  1 -1 -1  1  1 -1  1 -1 -1  1;]; % Walsh matrix

opndiff_walsh=(1/m)*W*Dstar*W;     % operational matrix for
                                   % differentiation
                                   % in Walsh domain
```

%%— -%%

```
WOTF=R*(inv((R*(eye(m,m)))+((L)*(opndiff_walsh/(T)))));
                                   % Walsh operational
                                   % transfer function
input_walsh=(1/m)*PWMM*W;          % input in Walsh domain
output_walsh=input_walsh*WOTF;     % output in Walsh domain
output_bpf=output_walsh*W          % output in BPF domain
```

%%— — — — PLOTTING LOAD CURRENT FOR Kd=0.25, 0.50 AND 0.75
%%IN WALSH DOMAIN

```
t=0:h:T
for i=1:3
    for j=1:m
            output_bpf1(1,j)=output_bpf(d(1,i),j);
    end
```

```
        output_bpf1
    hold on
    subplot(3,1,i)
    stairs(t,[output_bpf1 0],'k')   % plotting the coefficients
                                    % obtained
                                    % via Walsh function
    hold on
end


%%- - - - - - - - - LOAD CURRENT- - - - - - - - - - - -%%

for i=1:p(1,1)
    output_bpf1=0;
    output_bpf2=0;
    for j=1:m
        output_bpf1=[output_bpf(i,j)+output_bpf1];
        output_bpf2=[output_bpf(i,j)^2+output_bpf2];
    end
    avg_load_current(i)=(output_bpf1)/m;
    rms_load_current(i)=sqrt((output_bpf2)/m);
end

avg_load_current             % average load current obtained in
                             % Walsh domain
rms_load_current             % rms load current obtained in
                             % Walsh domain

%%- - - - - - - - - DEVICE CURRENT- - - - - - - - - - -%%

for i=1:p(1,1)
    for j=1:m
        output_bpf_dev(i,j)=output_bpf(i,j)*PWMM(i,j);
    end
end
for i=1:p(1,1)
    output_bpf1=0;
    output_bpf2=0;
    for j=1:m
        output_bpf1=[output_bpf_dev(i,j)+output_bpf1];
        output_bpf2=[output_bpf_dev(i,j)^2+output_bpf2];
    end
    avg_device_current(i)=(output_bpf1)/m;
    rms_device_current(i)=sqrt((output_bpf2)/m)
end

avg_device_current   % average device current obtained in Walsh
                     % domain
rms_device_current   % rms device current obtained in Walsh
                     % domain
```

```
%%- - - - - - - - - - FREEWHEELING CURRENT- - - - - - - - - - -%%

for i=1:p(1,1)
    for j=1:m
        output_bpf_diode(i,j)=output_bpf(i,j)*(1-PWMM(i,j));
    end
end

for i=1:p(1,1)
    output_bpf1=0;
    output_bpf2=0;
    for j=1:m
        output_bpf1=[output_bpf_diode(i,j)+output_bpf1];
        output_bpf2=[output_bpf_diode(i,j)^2+output_bpf2];
    end
    avg_diode_current(i)=(output_bpf1)/m;
    rms_diode_current(i)=sqrt((output_bpf2)/m)
end

avg_diode_current    % average diode current obtained in Walsh
                     % domain
rms_diode_current    % rms diode current obtained in Walsh
                     % domain

%%- - - - EXACT SOLUTION OF AVERAGE AND RMS CURRENTS- - - - %%

Tr=T/2                       % repitition rate of input waveform
Tl=L/R                       % load time constant
a=Tr/Tl
Kf=1-Kd                      % freewheeling cycle of chopper

for k=1:p(1,1)
    avg_device_exact(k)=Kd(k)-((1-exp(-(a*Kd(k))))/a)+((1-exp(-
    (a*Kd(k))))^2*(exp(-(a*Kf(k)))/(2*a)));    % equation (4.28)

    avg_diode_exact(k)=(1-exp(-(a*Kd(k))))*(1-exp(-
    (a*Kf(k))))*((2+exp(-a))/(2*a));% equation (4.30)

    avg_load_exact=avg_device_exact+ avg_diode_exact;
                                            % equation (4.31)

    A(k)=(1-exp(-(a*Kd(k))))/a;

    B(k)=(1-exp(-(2*a*Kd(k))))/(4*a);

    rms_device_exact(k)=sqrt(Kd(k)-(2*A(k))+(2*B(k))-
    (2*A(k)*B(k)*a*exp(-(a*Kf(k))))+((a^2)*(A(k)^2)*B(k)*
    exp(-(2*a*Kf(k))))+(a*(A(k)^2)*exp(-(a*Kf(k)))));
                                            % equation (4.33)
```

```
      rms_diode_exact(k)=(1/2)*(sqrt((a*A(k)^2)*(1-exp(-
      (2*a*Kf(k))))*(2+(2*exp(-a))+exp(-(2*a))))));
                                    % equation (4.34)

    rms_load_exact(k)=sqrt(rms_device_exact(k)^2+rms_diode_
    exact(k)^2);                    % equation (4.35)
end

avg_load_exact          % average load current obtained
                        % via exact solution
avg_device_exact        % average device current obtained via
                        % exact solution
avg_diode_exact         % average diode current obtained
                        % via exact solution
rms_load_exact          % rms load current obtained
                        % via exact solution
rms_device_exact        % rms device current obtained
                        % via exact solution
rms_diode_exact         % rms diode current obtained
                        % via exact solution


%%- - - - - PLOTTING LOAD CURRENT FOR Kd=0.25, 0.50 AND 0.75
                  % VIA EXACT SOLUTION- - - - - - - - - - -%%

Kd=0.25:0.25:3*0.25;       % selected duty cycles
                           % Kd=0.25,0.50,0.75
Kf=1-Kd;                   % corresponding freewheeling cycle of
                           % chopper
for k=1:3
      for i=1:81
          t(i)=(i-1)*0.001;
          if t(i)<(Kd(k)*Tr)
              func8(i)=1;             %%% func8=u(t-Kd*Tr)
          else
              func8(i)=0;
          end

          if t(i)<(Kd(k)*Tr)
              func1(i)=0;             %%% func1=u(t-Kd*Tr)
          else
              func1(i)=1;
          end

          if t(i)<(Tr+(Kd(k)*Tr))
              func2(i)=0;             %%% func2=u(t-Tr-Kd*Tr)
          else
              func2(i)=1;
          end
```

```
            if t(i)<(Tr)
                func5(i)=0;          %%% func5=u(t-Tr)
            else
                func5(i)=1;
        end
end


func=func5-func2 ;                   %%% func=u(t-Tr)-u(t-Tr-Kd*Tr)

for i=1:81
    t(i)=0.001*(i-1);
    func3(i)=exp(-(R*(t(i)-(Kd(k)*Tr)))/L);
    func4(i)=exp(-(R*(t(i)-Tr-(Kd(k)*Tr)))/L);
end

If1=(1-exp(-(a*Kd(k))))*func3*diag(func1); %If1=((1-exp(-
                                   % (a*Kd)))*exp(-(R*(t-
                                   % (Kd*Tr))/L)))*u(t-(Kd*Tr))

If2=(1-exp(-(a*Kd(k))))*(exp(-a)+1)*func4*diag(func2);
                                   % If2=((1-exp(-a*Kd)))*(exp(-
        % a)+1)*exp(-(R*(t-Tr-(Kd*Tr))/L)))*u(t-Tr-(Kd*Tr))
If=If1+If2;          % freewheeling current from equation (4.26)

for i=1:81
    t(i)=0.001*(i-1);
    func6(i)=(1-exp(-(R*t(i))/L));
    func7(i)=((1-exp(-(a*Kd(k))))*(exp(-(a*Kf(k)))))
    *(exp(-(R*(t(i)-Tr))/L)))+(1-exp(-(R*(t(i)-Tr))/L));
end

Is1=func6*diag(func8); % Is1=((1-exp(-(R*t)/L))*u(t))

Is2=func7*diag(func); % Is2=((1-exp(-(a*Kd)))*(exp-
                      % (a*Kf))*(exp(-(R*(t-
                      % Tr))/L))+(1-exp(-(R*(t-Tr))/L)))*u(t-Tr)

Is=Is1+Is2;          % device current from equation (4.25)

Il=If+Is;            % load current

hold on
subplot(3,1,k)
plot(t,Il,'k')
hold on
end
```

B.21 Program to Determine Average and rms Values of Load Current, Device Current, and Diode Current for a Given Input Voltage to an R–L Load (Example 4.1)

```
clc
clear all
format short
V=100           % input voltage
R=2.2           % load resistance
L=0.0002        % load inductance
KD=0.75         % duty cycle of chopper
Ton=0.00005     % on time of the switch
TT=Ton/KD       % Time interval for single on and off operation
Toff=(1-KD)*TT  % off time of the switch

%%— - - - - - - - - - - - - - - - - - - - - - - - - - - - - - -%%

A=[1 1 1 0]    % normalized input voltage in the
               % time interval 0-TT for KD=0.75

repitition_rate=64 % no. of switching on and off i.e.,
                                      % repitition rate
n=size(A);
T=repitition_rate*TT              % Total time considered
m=(n(1,2))*repitition_rate % no. of segments
h=T/m                            % length of each interval

%%— - - - - PULSE WIDTH MODULATION MATRIX FORMATION— - - - -%%

P=zeros(1,m);
for i=1:n(1,2)
    P(1,i)=P(1,i)+A(1,i);
end
k=size(A);
B=[zeros(1,k(1,2)) A];

for j=1:(m-n(1,2))/n(1,2)
    for i=n(1,2):(n(1,2))+4
        P(1,i)=P(1,i)+B(1,i);
end
k=size(B);
B=[zeros(1,k(1,2)) A];
n=n+4;
end
P;
PWMM=V*P                      % pulse width modulation matrix
```

```
%%- - - - - - - - - - - - - - - - - - - - - - - - - - - - - - -%%

QQ=eye(m-1,m-1);
p=zeros(m-1,1);
q=zeros(1,m);
Q=[p QQ;q];

opn_int_bpf=(T/m)*(1/2)*[eye(m,m)+Q]*([eye(m,m)-Q]^(-1));
                             % operational matrix for
                        % integration in block pulse domain
opn_diff_bpf=inv(opn_int_bpf);   % operational matrix for
                                 % differentiation in
                                 % block pulse domain

BPOTF=R*(inv((R*(eye(m,m)))+((L)*(opn_diff_bpf))));  % block
                                   % pulse operational
                                   % transfer function

output_bpf=(1/R)*PWMM*BPOTF;  % BPF coefficient of the output
                                          % load current

%%- - - - - - - - - - - - - - - - - - - - - - - - - - - - - -%%

t=0:h:T;
for i=1:1
    for j=1:m
        output_bpf1(j)=[output_bpf(i,j)];
    end
    stairs(t,[output_bpf1 0],'k')
    hold on
end

%%- - - - - - - - - - LOAD CURRENT- - - - - - - - - - - -%%

output_bpf1=0;
output_bpf2=0;
for j=1:m
    output_bpf1=[output_bpf(1,j)+output_bpf1];
    output_bpf2=[output_bpf(1,j)^2+output_bpf2];
end

avg_load_current=(output_bpf1)/m;

rms_load_current=sqrt((output_bpf2)/m);

avg_load_current                         % average load current
rms_load_current                         % rms load current
```

```
%%- - - - - - - - - - - DEVICE CURRENT- - - - - - - - - - -%%

for j=1:m
    output_bpf1(j)=[output_bpf(1,j)];
    output_bpf2(j)=[output_bpf(1,j)^2];
    P1(j)=[P(1,j)];
end

avg_device_current=(output_bpf1*transpose(P1))/m;

rms_device_current=sqrt((output_bpf2*transpose(P1))/m);

avg_device_current               % average device current
rms_device_current               % rms device current

%%- - - - - - - - - - FREEWHEELING CURRENT- - - - - - - - - -%%

for j=1:m
output_bpf1(j)=[output_bpf(1,j)];
output_bpf2(j)=[output_bpf(1,j)^2];
P1(j)=1-[P(1,j)];
end

avg_diode_current=(output_bpf1*transpose(P1))/m;

rms_diode_current=sqrt((output_bpf2*transpose(P1))/m);

avg_diode_current                % average diode current
rms_diode_current                % rms diode current
```

B.22 Program to Determine Average and rms Values of Load Current, Device Current, and Diode Current for a Given Input Voltage to an RL Load (Example 4.2)

```
clc
clear all
format short

V=220              % input voltage
R=5                % load resistance
L=7.5/(10^3)       % load inductance
f=1000        % frequency
TT=1/f        % Time interval for single on and off operation
KD=0.5        % duty cycle of chopper
```

```
%%— — — — — — — — — — — — — — — — — — — — — — — — — -%%

A=[1 1 0 0]   % normalized input voltage in the
                      % time interval 0-TT for KD=0.5

repitition_rate=64 % no. of switching on and
                          % off i.e., repitition rate

n=size(A);
T=repitition_rate*TT                % Total time considered
m=(n(1,2))*repitition_rate          % no. of segments
h=T/m                               % length of each interval

%%— — — — - PULSE WIDTH MODULATION MATRIX FORMATION— — — — -%%

P=zeros(1,m);
for i=1:n(1,2)
    P(1,i)=P(1,i)+A(1,i);
end
k=size(A);
B=[zeros(1,k(1,2)) A];

for j=1:(m-n(1,2))/n(1,2)
    for i=n(1,2):(n(1,2))+4
        P(1,i)=P(1,i)+B(1,i);
    end
    k=size(B);
    B=[zeros(1,k(1,2)) A];
    n=n+4;
end
P;                       % normalized input voltage in the
                                     % time interval 0-T
PWMM=V*P                 % pulse width modulation matrix

%%— — — — — — — — — — — — — — — — — — — — — — — — — -%%

QQ=eye(m-1,m-1);
p=zeros(m-1,1);
q=zeros(1,m);
Q=[p QQ;q];
opn_int_bpf=(T/m)*(1/2)*[eye(m,m)+Q]*([eye(m,m)-Q]^(-1));
                              % operational matrix for
                              % integration in block pulse domain
opn_diff_bpf=inv(opn_int_bpf); % operational matrix for
                                          % differentiation in
                              % block pulse domain
BPOTF=R*(inv((R*(eye(m,m)))+((L)*(opn_diff_bpf))));
            % block pulse operational transfer function
output_bpf=(1/R)*PWMM*BPOTF; % BPF coefficient of the output
                                          % load current
```

```
%%- - - - - - - - - - - - - - - - - - - - - - - - - - - - - -%%
t=0:h:T;
for i=1:1
    for j=1:m
        output_bpf1(j)=[output_bpf(i,j)];
    end
    stairs(t,[output_bpf1 0],'k')
    hold on
end

%%- - - - - - - - - - LOAD CURRENT- - - - - - - - - - - -%%
output_bpf1=0;
output_bpf2=0;
for j=1:m
    output_bpf1=[output_bpf(1,j)+output_bpf1];
    output_bpf2=[output_bpf(1,j)^2+output_bpf2];
end
avg_load_current=(output_bpf1)/m;
rms_load_current=sqrt((output_bpf2)/m);

avg_load_current                    % average load current
rms_load_current                    % rms load current

%%- - - - - - - - - - DEVICE CURRENT- - - - - - - - - - - -%%
for j=1:m
    output_bpf1(j)=[output_bpf(1,j)];
    output_bpf2(j)=[output_bpf(1,j)^2];
    P1(j)=[P(1,j)];
end
avg_device_current=(output_bpf1*transpose(P1))/m;
rms_device_current=sqrt((output_bpf2*transpose(P1))/m);

avg_device_current                  % average device current
rms_device_current                  % rms device current

%%- - - - - - - - - FREEWHEELING CURRENT- - - - - - - - - -%%
for j=1:m
    output_bpf1(j)=[output_bpf(1,j)];
    output_bpf2(j)=[output_bpf(1,j)^2];
    P1(j)=1-[P(1,j)];
end
avg_diode_current=(output_bpf1*transpose(P1))/m;
rms_diode_current=sqrt((output_bpf2*transpose(P1))/m);

avg_diode_current                   % average diode current
rms_diode_current                   % rms diode current
```

B.23 Variation of Normalized Average and rms Currents for an Inductive Load in a Half Controlled Rectifier with Triggering Angle α = 0°– 180° for *m* = 16 (Section 5.3, Figure 5.4)

```
clc
clear all
format long

V=1;            % normalized voltage
f=50;           % frequency of input
R=10            % load resistance
T=1/f           % time period for one full cycle
LL=[0.031831; 0.05513; 0.118795]; % different load inductance

%%- - - - - - - - - - - EXACT SOLUTION- - - - - - - - - - - -%%
    %%- - SOLUTION OF ANGLE BETA BY NEWTON RAPHSON METHOD- -%%

p=size(LL)
q=180/10
for k=1:p(1)
    L=LL(k)                     % select any one inductance
    XL=(2*pi*f*L);              % inductive reactance
    Z=sqrt(R^2+XL^2)            % load impedance
    phi_deg=atand(XL/R)         % phase angle

    phi_rad=(pi*phi_deg)/180;

for j=1:q
alfa_deg=10*(j-1);          % different triggering angle
                            % alfa=0 to 180 degree
alfa_rad=(alfa_deg*pi)/180;
initial_rad=0.011*2*pi*f; % initial value selection to
                    % obtain beta by newton raphson method

        for i=1:18
                val=initial_rad;

                a=(sin(val-phi_rad)-(sin(alfa_rad-phi_
rad))*exp((R/XL)*(alfa_rad-val))); % load current

                adiff=(cos(val-phi_rad)+(R*sin(alfa_rad-phi_
rad)*exp((R*(alfa_rad-val))/XL))/XL); % differentiation of
                                        % load current

                val1=a/adiff;
                initial_rad=val-val1;
        end
```

```
        beta_rad=initial_rad;
        beta_deg=(beta_rad*180)/pi        % angle beta
        gamma_deg=beta_deg-alfa_deg       % conduction angle gamma
        gamma_rad=beta_rad-alfa_rad;
        extc_deg=beta_deg-180;            % extinction angle

        %%— — — — — EXACT SOLUTION— — — — — %%

Iav1(j)=(1-cos(gamma_rad))*(cos(alfa_rad-phi_rad));
Iav2(j)=(sin(gamma_rad)-tan(phi_rad))*(sin(alfa_rad-phi_rad));
Iav3(j)=(sin(alfa_rad-phi_rad))*(tan(phi_rad))*exp(-cot(phi_
rad)*(gamma_rad));
Iav(j)=((V)/(2*pi))*(Iav1(j)+Iav2(j)+Iav3(j)); % exact
                                               % solution of average
                                               % load current


Irms2(j)=gamma_rad/2-sin(2*alfa_rad+2*gamma_rad-2*phi_rad)/4;
Irms3(j)=sin(2*alfa_rad-2*phi_rad)/4-sin(2*phi_rad-2*gamma_
rad)/64;
Irms4(j)=sin(2*gamma_rad-2*phi_rad)/64;
Irms5(j)=tan(phi_rad)/4-cot(phi_rad)/(cot(phi_rad)^2+1);
Irms6(j)=tan(phi_rad)/(4*exp(2*gamma_rad*cot(phi_rad)));
Irms7(j)=(cos(2*alfa_rad-2*phi_rad)*tan(phi_rad))/4;
Irms8(j)=sin(2*alfa_rad-2*phi_rad)/(cot(phi_rad)^2+1);
Irms9(j)=sin(2*alfa_rad+gamma_rad-2*phi_rad)/(exp(gamma_
rad*cot(phi_rad))*(cot(phi_rad)^2+1));
Irms10(j)=sin(gamma_rad)/(exp(gamma_rad*cot(phi_
rad))*(cot(phi_rad)^2+1));
Irms11(j)=sin(gamma_rad-2*phi_rad)/(8*exp(gamma_rad*cot(phi_
rad))*(cot(phi_rad)^2+1));
Irms12(j)=(cos(2*alfa_rad-2*phi_rad)*cot(phi_rad))/(cot(phi_
rad)^2+1);
Irms13(j)=(cos(2*alfa_rad-2*phi_rad)*tan(phi_rad))/
(4*exp(2*gamma_rad*cot(phi_rad)));
Irms14(j)=sin(2*phi_rad-gamma_rad)/(8*exp(gamma_rad*cot(phi_
rad))*(cot(phi_rad)^2+1));
Irms15(j)=(cot(phi_rad)*cos(gamma_rad-2*phi_rad))/
(16*exp(gamma_rad*cot(phi_rad))*(cot(phi_rad)^2+1));
Irms16(j)=(cot(phi_rad)*cos(gamma_rad+2*phi_rad))/
(16*exp(gamma_rad*cot(phi_rad))*(cot(phi_rad)^2+1));
Irms17(j)=(cos(-gamma_rad-2*phi_rad)*cot(phi_rad))/
(16*exp(gamma_rad*cot(phi_rad))*(cot(phi_rad)^2+1));
Irms18(j)=(cos(2*phi_rad-gamma_rad)*cot(phi_rad))/
(16*exp(gamma_rad*cot(phi_rad))*(cot(phi_rad)^2+1));
Irms19(j)=(cos(2*alfa_rad+gamma_rad-2*phi_rad)*cot(phi_rad))/
(exp(gamma_rad*cot(phi_rad))*(cot(phi_rad)^2+1));
Irms20(j)=(cos(gamma_rad)*cot(phi_rad))/(exp(gamma_
rad*cot(phi_rad))*(cot(phi_rad)^2+1));
Irms21(j)=Irms2(j)+Irms3(j)-Irms4(j)+Irms5(j);
```

```
Irms22(j)=-Irms6(j)-Irms7(j)-Irms8(j);
Irms23(j)=Irms9(j)-Irms10(j)+Irms11(j)+Irms12(j);
Irms24(j)=Irms13(j)+Irms14(j)-Irms15(j)-Irms16(j);
Irms25(j)=Irms17(j)+Irms18(j)-Irms19(j)+Irms20(j);

Irms1(j)=(V^2*(Irms21(j)+Irms22(j)+Irms23(j)+Irms24(j)+Irms
25(j)));

Irms(j)=sqrt((1/(2*pi))*Irms1(j));        % exact solution of rms
                                          % load current


%%- - - - - - - - - - BPF SOLUTION- - - - - - - - - - - - -%%


syms t
m=16                    % no. of Walsh components
h=T/m;                  % length of each interval

func=sin(2*pi*f*t)          % normalised voltage

    for i=1:m
        x1(i)=(1/h)*double(int(func,((i-1)*h),(i*h)));
            % BPF coefficients of input sinusoidal voltage
        bpfcoeff(1,i)=x1(i);
    end

    bpfcoeff            % BPF coefficients of normalised voltage

    x=((alfa_deg*m)/360)-fix((alfa_deg*m)/360);

    xx=fix((alfa_deg*m)/360);

    y=((beta_deg*m)/360)-fix((beta_deg*m)/360);

    yy=fix((beta_deg*m)/360);

    bpfcoeff(xx+1)=(1-x)*bpfcoeff(xx+1);    % area preservation
                                  % for (xx+1)th coefficient
    bpfcoeff(yy+1)=y*bpfcoeff(yy+1);        % area preservation
                                  % for (yy+1)th coefficient
    for i=1:xx
      bpfcoeff(i)=bpfcoeff(i)-bpfcoeff(i);
    end
    for i=yy+2:m
      bpfcoeff(i)=bpfcoeff(i)-bpfcoeff(i);
    end
    bpfcoeff        % coefficients of input voltage waveform for
                            % individual triggering angle alfa
```

```
        for i=1:m
            PCM(j,i)=bpfcoeff(1,i);
        end
    end
end

PCM                                  % phase control matrix

%%- - - - - - OPERATIONAL MATRIX IN WALSH DOMAIN- - - - - -%%

for i=1:16
    for j=1:16
        if i==j
            Dstar(i,j)=4*m*(1/2);
        elseif i>j
            Dstar(i,j)=0;
        else
            Dstar(i,j)=4*m*(-1)^(i+j);
        end
    end
end

Dstar;                          % equation (3.66)

    W=[1   1   1   1   1   1   1   1   1   1   1   1   1   1   1   1;
       1   1   1   1   1   1   1   1  -1  -1  -1  -1  -1  -1  -1  -1;
       1   1   1   1  -1  -1  -1  -1   1   1   1   1  -1  -1  -1  -1;
       1   1   1   1  -1  -1  -1  -1  -1  -1  -1  -1   1   1   1   1;
       1   1  -1  -1   1   1  -1  -1   1   1  -1  -1   1   1  -1  -1;
       1   1  -1  -1   1   1  -1  -1  -1  -1   1   1  -1  -1   1   1;
       1   1  -1  -1  -1  -1   1   1   1   1  -1  -1  -1  -1   1   1;
       1   1  -1  -1  -1  -1   1   1  -1  -1   1   1   1   1  -1  -1;
       1  -1   1  -1   1  -1   1  -1   1  -1   1  -1   1  -1   1  -1;
       1  -1   1  -1   1  -1   1  -1  -1   1  -1   1  -1   1  -1   1;
       1  -1   1  -1  -1   1  -1   1   1  -1   1  -1  -1   1  -1   1;
       1  -1   1  -1  -1   1  -1   1  -1   1  -1   1   1  -1   1  -1;
       1  -1  -1   1   1  -1  -1   1   1  -1  -1   1   1  -1  -1   1;
       1  -1  -1   1   1  -1  -1   1  -1   1   1  -1  -1   1   1  -1;
       1  -1  -1   1  -1   1   1  -1   1  -1  -1   1  -1   1   1  -1;
       1  -1  -1   1  -1   1   1  -1  -1   1   1  -1   1  -1  -1   1;];
                                        % Walsh matrix

    opnmatrix_walshdiff=(1/m)*W*Dstar*W;    % operational
                                            % matrix for
                              % differentiation in Walsh domain

%%- - - - - - - - - - - - - - - - - - - - - - - - - - - -%%

opnmatrix_walsh=inv(opnmatrix_walshdiff);
                              % operational matrix for
                              % integration in Walsh domain
```

```
      WOTF=(Z/(f*L))*inv(((R/(f*L))*eye(m,m))+(opnmatrix_
walshdiff));
                      % Walsh operational transfer function

%%- - - - - - - - - - - - - - - - - - - - - - - - - - - - - -%%

      input_walsh=(1/m)*PCM*W;          % system input in
                                        % Walsh domain
      output_walsh=input_walsh*WOTF;    % system output in
                                        % Walsh domain
      output_bpf=output_walsh*W         % system output in BPF
domain

%%- - - - SOLUTION OF AVERAGE AND R. M. S. LOAD CURRENT IN
WALSH DOMAIN- - - %%

      for i=1:q
          total=0;
          for j=1:m
              total=output_bpf(i,j)+total;
          end
          total1(i)=total ;
          avg_load_current(i)=total1(i)/m;
      end

      avg_load_current                  % average load current
                                        % in Walsh domain

      for i=1:q
          total2=0;
          for j=1:m
              total2=(output_bpf(i,j))^2+total2;
          end
          total3(i)=total2 ;
          rms_load_current(i)=sqrt(total3(i)/m);   % rms
                                        % load current
                                        % in Walsh domain
      end
      rms_load_current

%%- - - - - - - FUNCTION PLOTTING- - - - - - - - - %%

      for i=1:q
          alfa_deg(i)=10*(i-1);
      end
      subplot(2,1,1)
      plot(alfa_deg,Iav,'b') % plot of average load
                                        % current vs
              % triggering angle obtained by exact solution
      hold on
```

```
        subplot(2,1,1)
        plot(alfa_deg,avg_load_current,'k*')% plot of average
                                    % load current vs
            % triggering angle solved using Walsh function
        hold on
            subplot(2,1,2)
        plot(alfa_deg,Irms,'k*') % plot of rms load current vs
                                        % triggering angle
                        % obtained by exact solution
        hold on
        subplot(2,1,2)
        plot(alfa_deg,rms_load_current,'r')   % plot of rms load
                                        % current vs
            % triggering angle solved using Walsh function
        hold on
end
```

B.24 Solution of Average and rms Currents in a Half Controlled Rectifier with Triggering Angle α = 30°/120° for *m* = 16 (Example 5.1)

```
clc
clear all
format long

V=400                   % input voltage
f=50;                   % frequency of input
R=25                    % load resistance
T=1/f                   % time period for one full cycle
L=150/1000              % load inductance

%%- - - - - - - - - EXACT SOLUTION- - - - - - - - - - - -%%
%%- - SOLUTION OF ANGLE BETA BY NEWTON RAPHSON METHOD- -%%

XL=(2*pi*f*L);          % inductive reactance
Z=sqrt(R^2+XL^2)        % load impedance

phi_deg=atand(XL/R)     % phase angle
phi_rad=(pi*phi_deg)/180;

alfa_deg=30         % triggering angle alfa=30 degree (change
                    % triggering angle to 120 degree)
alfa_rad=(alfa_deg*pi)/180;
initial_rad=0.011*2*pi*f;        % initial value selection to
                                % obtain beta
                                % by newton raphson method
```

```
for i=1:18
    val=initial_rad;
    a=(sin(val-phi_rad)-(sin(alfa_rad-phi_rad))*exp((R/
    XL)*(alfa_rad-val))); % load current
    adiff=(cos(val-phi_rad)+(R*sin(alfa_rad-phi_
    rad)*exp((R*(alfa_rad-val))/XL))/XL); % differentiation of
                                          % load current
    val1=a/adiff;
    initial_rad=val-val1;
end


beta_rad=initial_rad;
beta_deg=(beta_rad*180)/pi          % angle beta
gamma_deg=beta_deg-alfa_deg         % conduction angle gamma
gamma_rad=(pi*gamma_deg)/180;
extc_deg=beta_deg-180;              % extinction angle


       %%- - - - - EXACT SOLUTION- - - - - %%

Iav=((V)/(2*pi*Z))*((1-cos(gamma_rad))*(cos(alfa_rad-phi_
rad))+(sin(gamma_rad)-tan(phi_rad))*(sin(alfa_rad-phi_
rad))+(sin(alfa_rad-phi_rad))*(tan(phi_rad))*exp(-cot(phi_
rad)*(gamma_rad)))


       %%- - - - - - - FUNTION PLOTTING- - - - - - - - - %%

alfa_time=(T/360)*alfa_deg              % triggering time
beta_time=(T/360)*beta_deg              % extinction time

for i=1:101
    t(i)=(i-1)*0.0002;
    if t<alfa_time
            func2(i)=0       %%% func2=u(t-2h)
    else
            func2(i)=1
    end
    if t<beta_time
            func4(i)=1       %%% func4=u(t-2h)
    else
            func4(i)=0
    end
end


for i=1:101
    k(i)=0.0002*(i-1);
    func1(i)=(V/Z)*(sin(2*pi*f*k(i)-phi_rad)-(sin(alfa_rad-
    phi_rad)*exp(cot(phi_rad)*(alfa_rad-(2*pi*f*k(i))))));
    end
```

```
func3=(func1*diag(func2))*diag(func4)     % exact solution
t=0:T/100:T
hold on
plot(t,func3,'r')

%%- - - - - - - - - WALSH DOMAIN SOLUTION- - - - - - - - - %%

syms t
m=16                                      % no. of Walsh component
h=T/m;                                    % length of each interval

func=sin(2*pi*f*t)                        % normalized input voltage
for i=1:m
    x1(i)=(1/h)*double(int(func,((i-1)*h),(i*h)));
    bpfcoeff1(1,i)=x1(i);                 % BPF coefficients of input
                                          % sinusoidal voltage
end

x=((alfa_deg*m)/360)-fix((alfa_deg*m)/360);
xx=fix((alfa_deg*m)/360);
y=((beta_deg*m)/360)-fix((beta_deg*m)/360);
yy=fix((beta_deg*m)/360);

bpfcoeff(xx+1)=(1-x)*bpfcoeff(xx+1);      % area preservation for
                                          % (xx+1)th coefficient
bpfcoeff(yy+1)=y*bpfcoeff(yy+1);        % area preservation for
                                          % (yy+1)th coefficient

for i=1:xx
    bpfcoeff(i)=bpfcoeff(i)-bpfcoeff(i);
end

for i=yy+2:m
        bpfcoeff(i)=bpfcoeff(i)-bpfcoeff(i);
end

bpfcoeff                       % coefficients of input voltage
                               % waveform for triggering angle alfa

PCM=bpfcoeff                   % phase control matrix

%%- - - - - OPERATIONAL MATRIX IN WALSH DOMAIN- - - - - -%%

for i=1:m
    for j=1:m
        if i==j
            Dstar(i,j)=4*m*(1/2);
        elseif i>j
            Dstar(i,j)=0;
        else
```

```
            Dstar(i,j)=4*m*(-1)^(i+j);

        end
    end
end

Dstar;                              % equation (3.66)

W=[1 1 1 1 1 1 1 1 1 1 1 1 1 1 1 1;
   1 1 1 1 1 1 1 1 -1 -1 -1 -1 -1 -1 -1 -1;
   1 1 1 1 -1 -1 -1 -1 1 1 1 1 -1 -1 -1 -1;
   1 1 1 1 -1 -1 -1 -1 -1 -1 -1 -1 1 1 1 1;
   1 1 -1 -1 1 1 -1 -1 1 1 -1 -1 1 1 -1 -1;
   1 1 -1 -1 1 1 -1 -1 -1 -1 1 1 -1 -1 1 1;
   1 1 -1 -1 -1 -1 1 1 1 1 -1 -1 -1 -1 1 1;
   1 1 -1 -1 -1 -1 1 1 -1 -1 1 1 1 1 -1 -1;
   1 -1 1 -1 1 -1 1 -1 1 -1 1 -1 1 -1 1 -1;
   1 -1 1 -1 1 -1 1 -1 -1 1 -1 1 -1 1 -1 1;
   1 -1 1 -1 -1 1 -1 1 1 -1 1 -1 -1 1 -1 1;
   1 -1 1 -1 -1 1 -1 1 -1 1 -1 1 1 -1 1 -1;
   1 -1 -1 1 1 -1 -1 1 1 -1 -1 1 1 -1 -1 1;
   1 -1 -1 1 1 -1 -1 1 -1 1 1 -1 -1 1 1 -1;
   1 -1 -1 1 -1 1 1 -1 1 -1 -1 1 -1 1 1 -1;
   1 -1 -1 1 -1 1 1 -1 -1 1 1 -1 1 -1 -1 1;] ; % Walsh matrix

opnmatrix_walshdiff=(1/m)*W*Dstar*W;    % operational matrix for
                        % differentiation in Walsh domain

%%- - - - - - - - - - - - - - - - - - - - - - - - - - -%%

opnmatrix_walsh=inv(opnmatrix_walshdiff); % operational
                                          % matrix for
                        % integration in Walsh domain

WOTF=(Z/(f*L))*inv(((R/(f*L))*eye(m,m))+(opnmatrix_
walshdiff));
                    % Walsh operational transfer function

%%- - - - - - - - - - - - - - - - - - - - - - - - - - -%%

input_walsh=(1/m)*PCM*W;        % system input in Walsh domain
output_walsh=input_walsh*WOTF; % system output in Walsh domain
output_bpf=((V)/Z)*output_walsh*W       % system output
                                        % in BPF domain

        %%- - - - - - - FUNCTION PLOTTING- - - - - - - - - %%

t=0:h:T
hold on
```

```
stairs(t,[output_bpf 0],'k')
hold on

%%- - - - SOLUTION OF AVERAGE AND R. M. S. LOAD CURRENT IN
WALSH DOMAIN- - - %%

total=0;
for i=1:m
    total=output_bpf(i)+total;
end
total;

avg_load_current=total/m              % average load current
                                      % in Walsh domain

total_rms=0;
for i=1:m
    total_rms=(output_bpf(i))^2+total_rms;
end

total_rms;

rms_load_current=sqrt(total_rms/m)    % rms load current in
                                      % Walsh domain
```

B.25 Solution of Average and rms Currents in a Half Controlled Rectifier with Triggering Angle $\alpha = 30°/120°$ for $m = 16$ Considering Two Cycles (Example 5.1)

```
clc
clear all
format long

V=400                          % input voltage
f=50;                          % frequency of input
R=25                           % load resistance
T=1/f                          % time period for one full cycle
L=150/1000                     % load inductance

%%- - - - - - - - - - EXACT SOLUTION- - - - - - - - - - - - -%%
    %%- - SOLUTION OF ANGLE BETA BY NEWTON RAPHSON METHOD- -%

XL=(2*pi*f*L);                 % inductive reactance
Z=sqrt(R^2+XL^2)               % load impedance
```

```
phi_deg=atand(XL/R)                    % phase angle
phi_rad=(pi*phi_deg)/180;

alfa_deg=30              % triggering angle alfa=30 degree
                         % (change triggering angle to 120 degree)
alfa_rad=(alfa_deg*pi)/180;
initial_rad=0.011*2*pi*f; % initial value selection to
                          % obtain beta
                          % by newton raphson method

for i=1:18
    val=initial_rad;
    a=(sin(val-phi_rad)-(sin(alfa_rad-phi_rad))*exp((R/
    XL)*(alfa_rad-val)));                        % load current
    adiff=(cos(val-phi_rad)+(R*sin(alfa_rad-phi_
    rad)*exp((R*(alfa_rad-val))/XL))/XL); % differentiation of
                                          % load current
    val1=a/adiff;
    initial_rad=val-val1;
end

beta_rad=initial_rad;
beta_deg=(beta_rad*180)/pi        % angle beta
gamma_deg=beta_deg-alfa_deg       % conduction angle gamma
gamma_rad=(pi*gamma_deg)/180;
extc_deg=beta_deg-180;            % extinction angle

         %%- - - - - EXACT SOLUTION- - - - - %%

Iav=((V)/(2*pi*Z))*((1-cos(gamma_rad))*(cos(alfa_rad-phi_
rad))+(sin(gamma_rad)-tan(phi_rad))*(sin(alfa_rad-phi_
rad))+(sin(alfa_rad-phi_rad))*(tan(phi_rad))*exp(-cot(phi_
rad)*(gamma_rad)))

         %%- - - - - - - FUNTION PLOTTING- - - - - - - - - %%

alfa_time=(T/360)*alfa_deg              % triggering time
beta_time=(T/360)*beta_deg              % extinction time

for i=1:101
    t(i)=(i-1)*0.0002;
    if t<alfa_time
            func2(i)=0       %%% func2=u(t-2h)
    else
            func2(i)=1
    end
    if t<beta_time
            func4(i)=1       %%% func4=u(t-2h)
```

```
     else
         func4(i)=0
     end
end

for i=1:101
    k(i)=0.0002*(i-1);
    func1(i)=(V/Z)*(sin(2*pi*f*k(i)-phi_rad)-(sin(alfa_rad-
    phi_rad)*exp(cot(phi_rad)*(alfa_rad-(2*pi*f*k(i))))));
end

func3=(func1*diag(func2))*diag(func4)    % exact solution
t=0:T/100:T
plot(t,func3,'r')
hold on
t=T:T/100:2*T
plot(t,func3,'r')

%%- - - - - - - - WALSH DOMAIN SOLUTION- - - - - - - - - %%

syms t
m=16                        % no. of Walsh coefficients
h=T/m;                      % length of each interval

func=sin(2*pi*f*t)          % normalised input voltage

for i=1:m
    x1(i)=(1/h)*double(int(func,((i-1)*h),(i*h)));
    bpfcoeff1(1,i)=x1(i);           % BPF coefficients of input
                                    % sinusoidal voltage
end

x=((alfa_deg*m)/360)-fix((alfa_deg*m)/360);
xx=fix((alfa_deg*m)/360);
y=((beta_deg*m)/360)-fix((beta_deg*m)/360);
yy=fix((beta_deg*m)/360);

bpfcoeff(xx+1)=(1-x)*bpfcoeff(xx+1);    % area preservation for
                                        % (xx+1)th coefficient
bpfcoeff(yy+1)=y*bpfcoeff(yy+1);        % area preservation for
                                        % (yy+1)th coefficient

for i=1:xx
    bpfcoeff(i)=bpfcoeff(i)-bpfcoeff(i);
end
for i=yy+2:m
    bpfcoeff(i)=bpfcoeff(i)-bpfcoeff(i);
end
```

```
bpfcoeff                          % coefficients of input voltage
                                  % waveform for triggering angle alfa

PCM=bpfcoeff                                      % phase control matrix

%%- - - - - - OPERATIONAL MATRIX IN WALSH DOMAIN- - - - - -%%

for i=1:m
    for j=1:m
        if i==j
            Dstar(i,j)=4*m*(1/2);
        elseif i>j
            Dstar(i,j)=0;
        else
            Dstar(i,j)=4*m*(-1)^(i+j);

        end
    end
end

Dstar;

W=[1  1  1  1  1  1  1  1  1  1  1  1  1  1  1  1;
   1  1  1  1  1  1  1  1 -1 -1 -1 -1 -1 -1 -1 -1;
   1  1  1  1 -1 -1 -1 -1  1  1  1  1 -1 -1 -1 -1;
   1  1  1  1 -1 -1 -1 -1 -1 -1 -1 -1  1  1  1  1;
   1  1 -1 -1  1  1 -1 -1  1  1 -1 -1  1  1 -1 -1;
   1  1 -1 -1  1  1 -1 -1 -1 -1  1  1 -1 -1  1  1;
   1  1 -1 -1 -1 -1  1  1  1  1 -1 -1 -1 -1  1  1;
   1  1 -1 -1 -1 -1  1  1 -1 -1  1  1  1  1 -1 -1;
   1 -1  1 -1  1 -1  1 -1  1 -1  1 -1  1 -1  1 -1;
   1 -1  1 -1  1 -1  1 -1 -1  1 -1  1 -1  1 -1  1;
   1 -1  1 -1 -1  1 -1  1  1 -1  1 -1 -1  1 -1  1;
   1 -1  1 -1 -1  1 -1  1 -1  1 -1  1  1 -1  1 -1;
   1 -1 -1  1  1 -1 -1  1  1 -1 -1  1  1 -1 -1  1;
   1 -1 -1  1  1 -1 -1  1 -1  1  1 -1 -1  1  1 -1;
   1 -1 -1  1 -1  1  1 -1  1 -1 -1  1 -1  1  1 -1;
   1 -1 -1  1 -1  1 -1 -1  1  1 -1  1 -1 -1  1;]; % Walsh matrix

opnmatrix_walshdiff=(1/m)*W*Dstar*W;  % operational matrix for
                                      % differentiation in Walsh domain

%%- - - - - - - - - - - - - - - - - - - - - - - - - - - - -%%

opnmatrix_walsh=inv(opnmatrix_walshdiff);     % operational
                                              % matrix for
                                      % integration in Walsh domain
```

```
WOTF=(Z/(f*L))*inv(((R/(f*L))*eye(m,m))+(opnmatrix_
walshdiff));
                    % Walsh operational transfer function

%%- - - - - - - - - - - - - - - - - - - - - - - - - - - - -%%

input_walsh=(1/m)*PCM*W;   % system input in Walsh domain
output_walsh=input_walsh*WOTF;          % system output in
                                        % Walsh domain
output_bpf=((V)/Z)*output_walsh*W       % system output in
                                        % BPF domain


        %%- - - - - - - FUNCTION PLOTTING- - - - - - - - - %%

t=0:h:T

hold on
stairs(t,[output_bpf 0],'k')

hold on
t=T:h:2*T
stairs(t,[output_bpf 0],'k')
hold on

%%- - - - SOLUTION OF AVERAGE AND R. M. S. LOAD CURRENT IN
                                   %% WALSH DOMAIN- - - %%

total=0;
for i=1:m
    total=output_bpf(i)+total;
end
total;

avg_load_current=total/m         % average load current
                                 % in Walsh domain

total_rms=0;

for i=1:m
    total_rms=(output_bpf(i))^2+total_rms;
end

total_rms;

rms_load_current=sqrt(total_rms/m)    % rms load current in
                                      % Walsh domain
```

B.26 Variation of Normalized Average and rms Currents for an Inductive Load in a Full-Wave Controlled Rectifier with Different Triggering Angles for *m* = 16 (Tables 5.1 and 5.2)

```
clc
clear all
format short

V=1                        % normalised voltage
f=50;                      % frequency of input
R=4.35                     % load resistance
T=1/f                      % time period for one full cycle
L=0.02                     % load inductance


%%- - - - - - - - - - EXACT SOLUTION - - - - - - - - - - - -%%

%%- - SOLUTION OF ANGLE BETA BY NEWTON RAPHSON METHOD- -%%

XL=(2*pi*f*L)              % inductive reactance
Z=sqrt(R^2+XL^2)           % load impedance
phi_deg=atand(XL/R)        % phase angle
phi_rad=(pi*phi_deg)/180;

alfa=[0.0 7.5 15.0 22.5 30.0 37.5 45.0 52.5 60.0 67.5 75.0
82.5 90.0 97.5 105.0 112.5 120.0];      % different
                                        % triggering angle alfa

kk=size(alfa)

for k=1:kk(2)
    alfa_deg=alfa(k);      % loop for selecting one triggering
                           % angle alfa
    alfa_rad=(alfa_deg*pi)/180;
    initial_rad=0.011*2*pi*f;    % initial value selection to
                                 % obtain beta by
                                 % newton raphson method
for i=1:18
        val=initial_rad;
        a=(sin(val-phi_rad)-(sin(alfa_rad-phi_rad))*exp((R/
        XL)*(alfa_rad-val)));                % load current
        adiff=(cos(val-phi_rad)+(R*sin(alfa_rad-phi_
        rad)*exp((R*(alfa_rad-val))/XL))/XL); % differentiation of
                                              % load current
        val1=a/adiff;
        initial_rad=val-val1;
end
```

```
beta_rad=initial_rad;
beta_deg=(beta_rad*180)/pi; % angle beta

if beta_deg>alfa_deg+180
beta_deg=alfa_deg+180;        % angle beta for
                              % continuous conduction
else beta_deg=beta_deg        % angle beta for
                              % discontinuous conduction
end

beta_rad=(beta_deg*pi)/180;
gamma_deg=beta_deg-alfa_deg;% conduction angle gamma
extc_deg=beta_deg-180;        % extinction angle
alfa2_deg=alfa_deg+180;       % triggering angle alfa for second
                              % half cycle
alfa2_rad=(alfa2_deg*pi)/180;

      %%- - - - - EXACT SOLUTION- - - - - %%

t1=((extc_deg*pi)/180)/(2*pi*f); % extinction time

t2=((alfa_deg*pi)/180)/(2*pi*f); % triggering time

w=2*pi*f;                           % angular frequency

a=-((1-exp(-((R*(T-t2))/L)))*sin((w*t2)-phi_rad));
b=((1-exp(-((R*((T/2)-t1))/L)))*sin((w*t1)+((w*T)/2)-phi_
rad));
c=((1-exp(-((R*((T/2)-t2))/L)))*sin((w*t2)+((w*T)/2)-phi_
rad));
d=(1/(w*T))*(cos(phi_rad)+cos((w*t2)-phi_rad)-
cos((w*t1)+((w*T)/2)-phi_rad)-cos((w*t2)+((w*T)/2)-phi_rad));

Iav(k)=((L/(R*T))*(a+b+c))+d;
Iav           % average load current IL1 by exact solution

%%- - - - - - - - - - - BPF SOLUTION - - - - - - - - - - - - -%%

syms t
m=16               % no. of Walsh components
h=T/m;             % length of each interval

    func=sin(2*pi*f*t)              % normalised input voltage
    for i=1:m
        x1(i)=(1/h)*double(int(func,((i-1)*h),(i*h)));
                    % BPF coefficients of input sinusoidal
                    % voltage
      bpfcoeff(1,i)=x1(i);
end
```

```
x=((alfa_deg*m)/360)-fix((alfa_deg*m)/360);
xx=fix((alfa_deg*m)/360);
x2=((alfa2_deg*m)/360)-fix((alfa2_deg*m)/360);
xx2=fix((alfa2_deg*m)/360);
y=((beta_deg*m)/360)-fix((beta_deg*m)/360);
yy=fix((beta_deg*m)/360);

bpfcoeff(xx+1)=(1-x)*bpfcoeff(xx+1);    % area preservation for
                                        % (xx+1)th
                     % coefficient for discontinuous conduction
bpf=y*bpfcoeff(yy+1);
bpf1=-((1-x2)*bpfcoeff(xx2+1));

if yy==xx2    % condition for continuous conduction
   bpfcoeff(xx2+1)=bpf1+bpf; % area preservation for (xx2+1)th
                     % coefficient for continuous conduction
else                  % condition for discontinuous conduction
bpfcoeff(yy+1)=y*bpfcoeff(yy+1);        % area preservation for
                                        % (yy+1)th coefficient
bpfcoeff(xx2+1)=-((1-x2)*bpfcoeff(xx2+1));% area
                                    % preservation for
                                    % (xx2+1)th coefficient
end
for i=1:xx
    bpfcoeff(i)=bpfcoeff(i)-bpfcoeff(i);
end

for i=xx2+2:m
    bpfcoeff(i)=-bpfcoeff(i);
end

for i=yy+2:xx2
    bpfcoeff(i)=bpfcoeff(i)-bpfcoeff(i);
end

bpfcoeff                      % BPF coefficients of input voltage

for i=1:m
    PCM(1,i)=bpfcoeff(1,i);      % phase control matrix
end

%%- - - - - - OPERATIONAL MATRIX IN WALSH DOMAIN - - - - - -%%

for i=1:16
    for j=1:16
        if i==j
            Dstar(i,j)=4*m*(1/2);
```

```
        elseif i>j
            Dstar(i,j)=0;
        else
            Dstar(i,j)=4*m*(-1)^(i+j);
        end
    end
end

Dstar;                          % equation (3.66)

W=[1  1  1  1  1  1  1  1  1  1  1  1  1  1  1  1;
   1  1  1  1  1  1  1  1 -1 -1 -1 -1 -1 -1 -1 -1;
   1  1  1  1 -1 -1 -1 -1  1  1  1  1 -1 -1 -1 -1;
   1  1  1  1 -1 -1 -1 -1 -1 -1 -1 -1  1  1  1  1;
   1  1 -1 -1  1  1 -1 -1  1  1 -1 -1  1  1 -1 -1;
   1  1 -1 -1  1  1 -1 -1 -1 -1  1  1 -1 -1  1  1;
   1  1 -1 -1 -1 -1  1  1  1  1 -1 -1 -1 -1  1  1;
   1  1 -1 -1 -1 -1  1  1 -1 -1  1  1  1  1 -1 -1;
   1 -1  1 -1  1 -1  1 -1  1 -1  1 -1  1 -1  1 -1;
   1 -1  1 -1  1 -1  1 -1 -1  1 -1  1 -1  1 -1  1;
   1 -1  1 -1 -1  1 -1  1  1 -1  1 -1 -1  1 -1  1;
   1 -1  1 -1 -1  1 -1  1 -1  1 -1  1  1 -1  1 -1;
   1 -1 -1  1  1 -1 -1  1  1 -1 -1  1  1 -1 -1  1;
   1 -1 -1  1  1 -1 -1  1 -1  1  1 -1 -1  1  1 -1;
   1 -1 -1  1 -1  1  1 -1  1 -1 -1  1 -1  1  1 -1;
   1 -1 -1  1 -1  1  1 -1 -1  1  1 -1  1 -1 -1  1;] ; % Walsh matrix
opnmatrix_walshdiff=(1/m)*W*Dstar*W; % operational matrix for
                        % differentiation in Walsh domain

%%- - - - - - - - - - - - - - - - - - - - - - - - - -%%

opnmatrix_walsh=inv(opnmatrix_walshdiff);       % operational
                                                % matrix for
                        % integration in Walsh domain
WOTF=(Z/(f*L))*inv(((R/(f*L))*eye(m,m))+(opnmatrix_
walshdiff));           % Walsh operational transfer function
%%- - - - - - - - - - - - - - - - - - - - - - - - - -%%

input_walsh=(1/m)*PCM*W;        % system input in Walsh domain
output_walsh=input_walsh*WOTF;  % system output in Walsh
                                % domain
output_bpf=output_walsh*W       % system output in BPF domain

%%- - SOLUTION OF AVERAGE AND R. M. S. LOAD CURRENT IN WALSH
%% DOMAIN - - %%

for i=1:1
    total=0;
    for j=1:m
```

```
        total=output_bpf(i,j)+total;
        end
        total1(i)=total ;
        avg_current(i)=total1(i)/(m);
end

avg_load_current(k)=avg_current          % average load current
                                         % in Walsh domain


Iav

for i=1:1
    total2=0;
    for j=1:m
    total2=(output_bpf(i,j))^2+total2;
    end
    total3(i)=total2 ;
    rms_current(i)=sqrt(total3(i)/m);
end

rms_load_current(k)=rms_current   % rms load current in
                                  % Walsh domain

end
```

B.27 Variation of Normalized Average and rms Currents for an Inductive Load in a Full-Wave Controlled Rectifier with Different Triggering Angles for *m* = 16 (Tables 5.3 and 5.4)

```
clc
clear all
format short

V=1                 % normalised voltage
f=50;               % frequency of input
R=4.35              % load resistance
T=1/f               % time period for one full cycle
L=0.02              % load inductance

%%- - - - - - - - - - EXACT SOLUTION- - - - - - - - - - - -%%

%%- - SOLUTION OF ANGLE BETA BY NEWTON RAPHSON METHOD- -%%

XL=(2*pi*f*L)       % inductive reactance
Z=sqrt(R^2+XL^2)    % load impedance
```

```
phi_deg=atand(XL/R) % phase angle

phi_rad=(pi*phi_deg)/180;

alfa=[0.0 7.5 15.0 22.5 30.0 37.5 45.0 52.5 60.0 67.5 75.0
82.5 90.0 97.5 105.0 112.5 120.0];          % different triggering
                                            % angle alfa

kk=size(alfa)
for k=1:kk(2)
    alfa_deg=alfa(k);               % loop for selecting one
    triggering
                                    % angle alfa
    alfa_rad=(alfa_deg*pi)/180;
    initial_rad=0.011*2*pi*f;       % initial value selection to
                                    % obtain
                        % beta by newton raphson method
    for i=1:18
        val=initial_rad;
        a=(sin(val-phi_rad)-(sin(alfa_rad-phi_rad))*exp((R/
        XL)*(alfa_rad-val)));              % load current
        adiff=(cos(val-phi_rad)+(R*sin(alfa_rad-phi_rad)
        *exp((R*(alfa_rad-val))/XL))/XL); % differentiation of
                                          % load current
        val1=a/adiff;
        initial_rad=val-val1;
    end

beta_rad=initial_rad;
beta_deg=(beta_rad*180)/pi          % angle beta

if beta_deg>alfa_deg+180
        beta_deg=alfa_deg+180;      % angle beta for
                                    % continuous conduction
else beta_deg=beta_deg       % angle beta for discontinuous
                             % conduction
end

beta_rad=(beta_deg*pi)/180;
gamma_deg=beta_deg-alfa_deg; % conduction angle gamma
extc_deg=beta_deg-180        % extinction angle
alfa2_deg=alfa_deg+180       % triggering angle alfa for second
                             % half cycle
alfa2_rad=(alfa2_deg*pi)/180;

        %%- - - - - EXACT SOLUTION- - - - - %%

t1=((extc_deg*pi)/180)/(2*pi*f); % extinction time
t2=((alfa_deg*pi)/180)/(2*pi*f); % triggering time
```

```
w=2*pi*f;                      % angular frequency

a=-((1-exp(-((R*T)/L)))*sin(phi_rad));

b=-((1-exp(-((R*(T-t1))/L)))*sin((w*t1)-phi_rad));

c=-((1-exp(-((R*(T-t2))/L)))*sin((w*t2)-phi_rad));

d=((1-exp(-((R*((T/2)-t1))/L)))*sin((w*t1)+((w*T)/2)-phi_
rad));

e=((1-exp(-((R*((T/2)-t2))/L)))*sin((w*t2)+((w*T)/2)-phi_
rad));

g=(1/(w*T))*(cos((w*t1)-phi_rad)+cos((w*t2)-phi_rad)-
cos((w*t1)+((w*T)/2)-phi_rad)-cos((w*t2)+((w*T)/2)-phi_rad));

Iav(k)=((L/(R*T))*(a+b+c+d+e))+g;

Iav            % average load current IL2 by exact solution

%%- - - - - - - - - - BPF SOLUTION- - - - - - - - - - - -%%

syms t
m=16
h=T/m;                         % length of each interval
func=sin(2*pi*f*t)             % normalised input voltage
for i=1:m
    x1(i)=(1/h)*double(int(func,((i-1)*h),(i*h)));
                % BPF coefficients of input sinusoidal voltage
    bpfcoeff(1,i)=x1(i);
end

x=((alfa_deg*m)/360)-fix((alfa_deg*m)/360)
xx=fix((alfa_deg*m)/360)
x2=((alfa2_deg*m)/360)-fix((alfa2_deg*m)/360)
xx2=fix((alfa2_deg*m)/360)

y=((beta_deg*m)/360)-fix((beta_deg*m)/360)
yy=fix((beta_deg*m)/360)
y1=((extc_deg*m)/360)-fix((extc_deg*m)/360)
yy1=fix((extc_deg*m)/360)

bpfcoeff(xx+1)=(1-x)*bpfcoeff(xx+1)     % area preservation for
                                        % (xx+1)th
                    % coefficient for discontinuous conduction
bpf=y*bpfcoeff(yy+1)

bpf1=-((1-x2)*bpfcoeff(xx2+1))

if yy==xx2 && yy1==xx      % condition for continuous
                           % conduction
```

```
         bpfcoeff(xx2+1)=bpf1+bpf % area preservation for
                                           % (xx2+1)th
                     % coefficient for continuous conduction
         bpfcoeff(xx+1)=bpf1+bpf % area preservation for
                                           % (xx+1)th
                     % coefficient for continuous conduction

else          % condition for discontinuous conduction
         bpfcoeff(yy1+1)=-(y1*bpfcoeff(yy1+1))    % area
                                 % preservation for
                                 % (yy1+1)th coefficient

         bpfcoeff(yy+1)=y*bpfcoeff(yy+1)      % area preservation for
                                 % (yy+1)th coefficient
         bpfcoeff(xx2+1)=-((1-x2)*bpfcoeff(xx2+1)) % area
                                 % preservation for
                                 % (xx2+1)th coefficient
end

for i=xx2+2:m
     bpfcoeff(i)=-bpfcoeff(i);
end

for i=1:yy1
     bpfcoeff(i)=-bpfcoeff(i);
end

for i=yy1+2:xx
     bpfcoeff(i)=bpfcoeff(i)-bpfcoeff(i);
end

for i=yy+2:xx2
     bpfcoeff(i)=bpfcoeff(i)-bpfcoeff(i);
end

bpfcoeff                      % BPF coefficients of input voltage

for i=1:m
     PCM(1,i)=bpfcoeff(1,i);               % phase control matrix
end

%%- - - - - - OPERATIONAL MATRIX IN WALSH DOMAIN- - - - - -%%

for i=1:16
     for j=1:16

             if i==j
                Dstar(i,j)=4*m*(1/2);
             elseif i>j
```

```
                    Dstar(i,j)=0;
                else
                    Dstar(i,j)=4*m*(-1)^(i+j);
                end
        end
end


Dstar;                          % equation (3.66)


W=[1  1  1  1  1  1  1  1  1  1  1  1  1  1  1  1;
   1  1  1  1  1  1  1  1 -1 -1 -1 -1 -1 -1 -1 -1;
   1  1  1  1 -1 -1 -1 -1  1  1  1  1 -1 -1 -1 -1;
   1  1  1  1 -1 -1 -1 -1 -1 -1 -1 -1  1  1  1  1;
   1  1 -1 -1  1  1 -1 -1  1  1 -1 -1  1  1 -1 -1;
   1  1 -1 -1  1  1 -1 -1 -1 -1  1  1 -1 -1  1  1;
   1  1 -1 -1 -1 -1  1  1  1  1 -1 -1 -1 -1  1  1;
   1  1 -1 -1 -1 -1  1  1 -1 -1  1  1  1  1 -1 -1;
   1 -1  1 -1  1 -1  1 -1  1 -1  1 -1  1 -1  1 -1;
   1 -1  1 -1  1 -1  1 -1 -1  1 -1  1 -1  1 -1  1;
   1 -1  1 -1 -1  1 -1  1  1 -1  1 -1 -1  1 -1  1;
   1 -1  1 -1 -1  1 -1  1 -1  1 -1  1  1 -1  1 -1;
   1 -1 -1  1  1 -1 -1  1  1 -1 -1  1  1 -1 -1  1;
   1 -1 -1  1  1 -1 -1  1 -1  1  1 -1 -1  1  1 -1;
   1 -1 -1  1 -1  1  1 -1  1 -1 -1  1 -1  1  1 -1;
   1 -1 -1  1 -1  1  1 -1 -1  1  1 -1  1 -1 -1  1;] ; % Walsh matrix


opnmatrix_walshdiff=(1/m)*W*Dstar*W; % operational matrix
                                     % for
                           % differentiation in Walsh domain


%%- - - - - - - - - - - - - - - - - - - - - - - - - - - -%%


opnmatrix_walsh=inv(opnmatrix_walshdiff);% operational
                                         % matrix for
                          % integration in Walsh domain


WOTF=(Z/(f*L))*inv(((R/(f*L))*eye(m,m))+(opnmatrix_
walshdiff));        % Walsh operational transfer function
%%- - - - - - - - - - - - - - - - - - - - - - - - - - - -%%


input_walsh=(1/m)*PCM*W;   % system input in Walsh domain
output_walsh=input_walsh*WOTF;   % system output in
                                 % Walsh domain
output_bpf=output_walsh*W         % system output in BPF domain
```

```
%%- - - - SOLUTION OF AVERAGE AND R. M. S. LOAD CURRENT IN
%% WALSH DOMAIN- - - %%

        for i=1:1
            total=0;
            for j=1:m
                total=output_bpf(i,j)+total;
            end
            total1(i)=total ;
            avg_current(i)=total1(i)/m;
        end

        avg_load_current(k)=avg_current % average load
                                        % current in Walsh domain

        for i=1:1
            total2=0;
            for j=1:m
                total2=(output_bpf(i,j))^2+total2;
            end
        total3(i)=total2 ;
        rms_current(i)=sqrt(total3(i)/m);
        end

rms_load_current(k)=rms_current % rms load current
                                % in Walsh domain

end
```

B.28 Computation of Normalized Average and rms Currents for Single-Pulse Modulation (Section 6.1.1.2, Tables 6.1 and 6.2)

```
clc
clear all
format short

V=1             % normalized voltage
R=4.35          % load resistance
L=0.02          % load inductance
T=0.02          % normalized time
f=50;           % frequency of input
m=16
h=T/m           % length of each interval
```

```
angle1=transpose([0 4.5 9.0 13.5 18 27 36 45 49.5 54 58.5 63
72 81])
                    % different angles to trigger the thyristor
p=size(angle1);

%%- - - - - - - - - - - - - - - - - - - - - - - - - - - - -%%

for i=1:p(1,1)
    angle=angle1(i,1);              % select each angle
    t1=((angle*pi)/180)/(2*pi*f);
                % equivalent time by which the pulse is advanced
    norm_t1(1,i)=t1/T;             % equivalent normalised time

%%- - - - - - - - - - - EXACT SOLUTION- - - - - - - - - -%%

Iav=(2/T)*(((T/2)-(2*t1))+(L/R)*((exp(-(R*((T/2)-t1)/L)))-
(exp(-((R*t1)/L))))));
                % exact solution of normalised average current

avg_current_exact(1,i)=Iav;      % average current

%%- - - - - - - - WALSH DOMAIN SOLUTION- - - - - - - - - - %%

xx=t1/h;
x1=xx-fix(xx);
x=1-x1;
q=((m/2)-(2*((fix(xx)))));
n=q-2;                   % no. of terms n

%%- - - - - - - - - - - - - - - - - - - - - - - - - - - -%%

if ((m/4)-(n/2))-1>0
    for j=1:((m/4)-(n/2))-1
    A(1,j)=V*0;           % first (m/4-n/2-1)terms of equation 6.7
    end
end

A(1,((m/4)-(n/2)))=V*x;    % (m/4-n/2)th term

B=A;

if n>1
        for j=1:n
            C(1,j)=V*1;       % next n terms of equation 6.7
            end
        D=[B C];
        C=0;
else D=B;
end
```

```
k=size(D);
D(1,k(2)+1)=V*x;              % next term of equation 6.7
if ((m/4)-(n/2))-1>0
       for j=1:((m/4)-(n/2))-1
       E(1,j)=V*0;       % last (m/4-n/2-1)zeros of equation 6.7
       end
       F=[D E];       % (m/2) terms of equation 6.7 are ready
else F=D;             % (m/2) terms of equation 6.7 are ready
end
bpfcoeff=[F -F];             % m terms of equation 6.7 are ready


for j=1:m
       SPMM(i,j)=bpfcoeff(1,j);
                        % single pulse-width modulation matrix for
                        % different angles
end


%%- - - - - - - - - - - - - - - - - - - - - - - - - - - - - - %%


for i=1:m
    for j=1:m
        if i==j
            Dstar(i,j)=4*m*(1/2);
        elseif i>j
            Dstar(i,j)=0;
        else
            Dstar(i,j)=4*m*(-1)^(i+j);
        end
    end
end
Dstar;                       % equation (3.66)

W=[1  1  1  1  1  1  1  1  1  1  1  1  1  1  1  1;
   1  1  1  1  1  1  1  1 -1 -1 -1 -1 -1 -1 -1 -1;
   1  1  1  1 -1 -1 -1 -1  1  1  1  1 -1 -1 -1 -1;
   1  1  1  1 -1 -1 -1 -1 -1 -1 -1 -1  1  1  1  1;
   1  1 -1 -1  1  1 -1 -1  1  1 -1 -1  1  1 -1 -1;
   1  1 -1 -1  1  1 -1 -1 -1 -1  1  1 -1 -1  1  1;
   1  1 -1 -1 -1 -1  1  1  1  1 -1 -1 -1 -1  1  1;
   1  1 -1 -1 -1 -1  1  1 -1 -1  1  1  1  1 -1 -1;
   1 -1  1 -1  1 -1  1 -1  1 -1  1 -1  1 -1  1 -1;
   1 -1  1 -1  1 -1  1 -1 -1  1 -1  1 -1  1 -1  1;
   1 -1  1 -1 -1  1 -1  1  1 -1  1 -1 -1  1 -1  1;
   1 -1  1 -1 -1  1 -1  1 -1  1 -1  1  1 -1  1 -1;
   1 -1 -1  1  1 -1 -1  1  1 -1 -1  1  1 -1 -1  1;
   1 -1 -1  1  1 -1 -1  1 -1  1  1 -1 -1  1  1 -1;
   1 -1 -1  1 -1  1  1 -1  1 -1 -1  1 -1  1  1 -1;
   1 -1 -1  1 -1  1  1 -1 -1  1  1 -1  1 -1 -1  1;];   % Walsh matrix
```

```
opndiff_walsh=(1/m)*W*Dstar*W;    % operational matrix for
                                  % differentiation
                                  % in Walsh domain
WOTF=R*(inv((R*eye(m,m))+(opndiff_walsh*(L/T))));    % Walsh
                                             % operational
                                  % transfer function
input_walsh=(1/m)*SPMM*W;         % system input in Walsh domain
output_walsh=input_walsh*WOTF;         % system output in
                                  % Walsh domain
output_bpf=output_walsh*W;        % system output in BPF domain
end


norm_t1
SPMM                  % single pulse width modulation matrix

%%- - - - - - - - - - - - - - - - - - - - - - - - - - - -%%

for i=1:p(1,1)
output_bpf1=0;
output_bpf2=0;
for j=1:m/2
       output_bpf1=[output_bpf(i,j)+output_bpf1];
       output_bpf2=[output_bpf(i,j)^2+output_bpf2];
end
avg_load_current(i)=(output_bpf1)/(m/2);         % average
                                        % load current
                                  % solved in Walsh domain
rms_load_current(i)=sqrt((output_bpf2)/(m/2)); % rms load
                                  % current solved
                                  % in Walsh domain
end

avg_current_exact
avg_load_current
rms_load_current
```

B.29 Computation of Normalized Average and rms Currents for Three-Phase Inverter Line to Phase Voltage (Section 6.2)

```
clc
clear all
format short

V=1             % normalized voltage
R=4.35          % load resistance
L=0.02          % load inductance
```

```
T=0.02          % normalized time
f=50;           % frequency of input
m=16
h=T/m           % length of each interval

%%- - - - - - - - - - EXACT SOLUTION- - - - - - - - - - - - %%

avg_current_exact=(2/(3*T))*(((2*T)/3)+((L/R)*(exp(-(R*T)/
(2*L))+exp(-(R*T)/(3*L))-exp(-(R*T)/(6*L))-1)));
                % exact solution of normalised average current

%%- - - - - - - - - - - WALSH DOMAIN SOLUTION- - - - - - - -%%

p=log(m)/log(2)

q=(1/6)*((2^p)-3-((-1)^p)) % value of q from equation 6.21

qdash=(1/2)*((2*q)-1+((-1)^p))              % value of qdash from
                                            % equation 6.21

%%- - - - - - - - - - - - - - - - - - - - - - - - - - - - - -%%

for j=1:q
    A(1,j)=V*(1/3)          % first q terms of equation 6.20
end

A(1,q+1)=(1/3)*((m/3)-q-qdash)          % (q+1)th term of
                                        % equation 6.20
for j=1:qdash
    B(1,j)=V*(2/3)          % next qdash terms of equation 6.20
end

C=[A B]

k=size(C)

C(1,k(1,2)+1)=(1/3)*((m/3)-q-qdash)

for j=1:q
    D(1,j)=V*(1/3)   % last q terms of equation 6.20
end

E=[C D]

SPMM=[E-E]              % single pulse width modulation matrix

%%- - - - - - - - - - - - - - - - - - - - - - - - - - - - -%%

for i=1:m
    for j=1:m
```

```
         if i==j
            Dstar(i,j)=4*m*(1/2);
         elseif i>j
            Dstar(i,j)=0;
         else
            Dstar(i,j)=4*m*(-1)^(i+j);
         end
     end
end
Dstar;                              % equation (3.66)

W=[1  1  1  1  1  1  1  1  1  1  1  1  1  1  1  1;
   1  1  1  1  1  1  1  1 -1 -1 -1 -1 -1 -1 -1 -1;
   1  1  1  1 -1 -1 -1 -1  1  1  1  1 -1 -1 -1 -1;
   1  1  1  1 -1 -1 -1 -1 -1 -1 -1 -1  1  1  1  1;
   1  1 -1 -1  1  1 -1 -1  1  1 -1 -1  1  1 -1 -1;
   1  1 -1 -1  1  1 -1 -1 -1 -1  1  1 -1 -1  1  1;
   1  1 -1 -1 -1 -1  1  1  1  1 -1 -1 -1 -1  1  1;
   1  1 -1 -1 -1 -1  1  1 -1 -1  1  1  1  1 -1 -1;
   1 -1  1 -1  1 -1  1 -1  1 -1  1 -1  1 -1  1 -1;
   1 -1  1 -1  1 -1  1 -1 -1  1 -1  1 -1  1 -1  1;
   1 -1  1 -1 -1  1 -1  1  1 -1  1 -1 -1  1 -1  1;
   1 -1  1 -1 -1  1 -1  1 -1  1 -1  1  1 -1  1 -1;
   1 -1 -1  1  1 -1 -1  1  1 -1 -1  1  1 -1 -1  1;
   1 -1 -1  1  1 -1 -1  1 -1  1  1 -1 -1  1  1 -1;
   1 -1 -1  1 -1  1  1 -1  1 -1 -1  1 -1  1  1 -1;
   1 -1 -1  1 -1  1  1 -1 -1  1  1 -1  1 -1 -1  1;];  % Walsh matrix

opndiff_walsh=(1/m)*W*Dstar*W;   % operational matrix for
                        % differentiation in Walsh domain
WOTF=R*(inv((R*eye(m,m))+(opndiff_walsh*(L/T))));
                  % Walsh operational transfer function

input_walsh=(1/m)*SPMM*W;        % system input in Walsh domain

output_walsh=input_walsh*WOTF;   % system output
                                 % in Walsh domain

output_bpf=output_walsh*W;       % system output in BPF domain

SPMM                  % single pulse width modulation matrix

output_bpf            % load current

%%- - - - - - - - - - - - - - - - - - - - - - - - - - - - -%%

output_bpf1=0;

output_bpf2=0;
```

```
for j=1:m/2
    output_bpf1=[output_bpf(1,j)+output_bpf1];
    output_bpf2=[output_bpf(1,j)^2+output_bpf2];
end

avg_load_current=(output_bpf1)/(m/2);
            % average load current solved in Walsh domain
rms_load_current=sqrt((output_bpf2)/(m/2));
                    % rms load current solved in Walsh domain

avg_current_exact
avg_load_current
rms_load_current
```

Index

Note: Locators "*f*" and "*t*" denote figures and tables in the text